Electronic Interpretation of Organic Chemistry

A Problems-Oriented Text

Electronic Interpretation of Organic Chemistry

A Problems-Oriented Text

Fredric M. Menger
and
Leon Mandell

Emory University
Atlanta, Georgia

PLENUM PRESS · NEW YORK AND LONDON

Library of Congress Cataloging in Publication Data

Menger, Fred M 1937-
  Electronic interpretation of organic reactions.

  Includes index.
  1. Chemical reactions. 2. Chemistry, Organic. I. Mandell, Leon, joint author.
II. Title.
QD502.M46   547′.1′39   79-21718

  ISBN-13:978-1-4684-3667-9   e-ISBN-13:978-1-4684-3665-5
  DOI: 10.1007/978-1-4684-3665-5

© 1980 Plenum Press, New York
Softcover reprint of the hardcover 1st edition 1980

A Division of Plenum Publishing Corporation
227 West 17th Street, New York, N.Y. 10011

Preface

Most standard texts in basic organic chemistry require the student to memorize dozens of organic reactions. This is certainly necessary to master the discipline. Unfortunately, most texts do not emphasize *why* these reactions occur and, just as important, why other reactions that might seem conceivable to the student do *not* occur. Without this understanding, students tend to forget what they have memorized soon after the course is over. It is the purpose of this book to familiarize the student with the principles governing organic reactivity and to provide a "feel" for organic chemistry that is impossible to secure by memory alone. Digesting the ideas in this book will, we hope, not only explain the common organic reactions but also allow the student to predict the products and by-products of reactions he has never seen before. Indeed, the creative student might even become capable of designing new reactions as might be required in a complex organic synthesis.

In Chapter 1, we cover the basic principles including bonding, nuclear charge, resonance effects, oxidation–reduction, etc. It is a brief discussion, but it nonetheless provides the basis for understanding reaction mechanisms that will be treated later on. We highly recommend that this material be reviewed and that the problems be worked at the end of the chapter. Answers are given to all problems.

In Chapter 2, reaction mechanisms are presented in an increasing order of difficulty. Most of the reactions possess more than one step and often entail several intermediates of varying stability. The text following each mechanism teaches how to assess the reasonableness of the proposed intermediates in reaction mechanisms. The text also discusses whether or not bond formation and bond breakage in the various steps are consistent with valid chemical principles.

Chapter 3 provides additional problems and answers. Many of the important organic intermediates (carbonium ions, carbanions, radicals, carbenes, benzyne, etc.) are encountered in these problems.

Finally, Chapter 4 deals with a molecular orbital approach to organic chemistry and includes problems and answers on the subject. The concepts are simple and nonmathematical. Together with the "electron pushing" method the student is provided with a qualitative, yet powerful, means of handling organic reactions.

Emory University   Fredric M. Menger
                                                            Leon Mandell

Contents

CHAPTER 1

Basic Principles

1.1. Bonding

It is assumed that the student using this book has been exposed to the rudiments of bonding and structure, so that the following information should serve mainly as a review of the basic principles necessary in understanding organic reaction mechanisms.

Covalent bonds form when two atoms share a pair of electrons; the presence of the negatively charged electrons between two positively charged nuclei holds the atoms together. Carbon, with four electrons in its outer shell, can combine with four hydrogens, each with one electron in the outer shell:

$$
\overset{\displaystyle \overset{H}{\cdot}}{\underset{\displaystyle \underset{H}{\cdot}}{H\cdot \cdot \overset{\cdot}{\underset{\cdot}{C}}\cdot \cdot H}} \rightarrow
\overset{\displaystyle H}{\underset{\displaystyle H}{H : \overset{\cdot\cdot}{\underset{\cdot\cdot}{C}} : H}} \quad \text{or} \quad
\overset{\displaystyle H}{\underset{\displaystyle H}{H - \overset{|}{\underset{|}{C}} - H}}
$$

Thus, carbon achieves a full octet of electrons (the maximum that the second shell can sustain). The carbon in methane is, of course, not planar, but tetrahedral, since it is sp^3 hybridized. Ethylene, on the other hand, is composed of two sp^2-hybridized carbons linked by a double bond. Note that one of the two carbon–carbon double bonds is characterized by "sideways" overlap between two *p* orbitals:

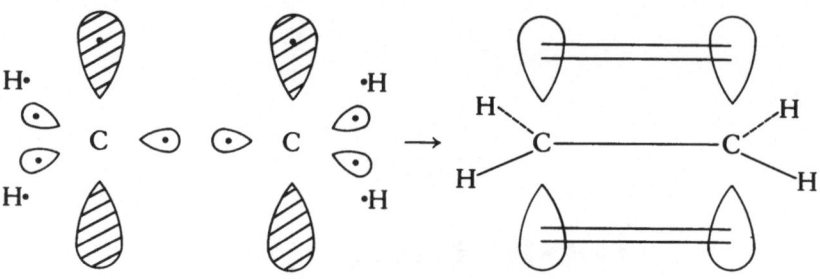

Sideways overlap, creating a so-called π bond, is, in general, less effective than the "head-on" overlap involved in the formation of the second carbon–carbon bond (i.e., the σ bond). Thus, the π bond represents a center of high electron density and reactivity. Ethylene will react, for example, with bromine thereby destroying the π bond:

$$CH_2{=}CH_2 + Br_2 \longrightarrow \underset{\underset{Br}{|}}{CH_2}{-}\underset{\underset{Br}{|}}{CH_2}$$

Acetylene possesses a triple bond comprised of two π bonds and a σ bond. Each carbon is *sp* hybridized and bears a total of eight bonding electrons.

$$H{-}C{\equiv}C{-}H$$

Oxygen has six outer electrons and therefore can bond to only two hydrogen atoms:

$$H{\cdot} \quad \cdot\ddot{\underset{\cdot}{O}}{:} \longrightarrow H{-}\underset{\underset{H}{|}}{\dot{C}}{:}$$

Note that there are two pairs of electrons on neutral oxygen that are not engaged in bonding. These unshared or nonbonding electrons play an important role in the chemistry of oxygen-containing organic compounds. Nitrogen with five outer electrons and chlorine with seven are trivalent and univalent, respectively:

$$
3\,\text{H}\cdot + \cdot\ddot{\text{N}}\!: \longrightarrow \text{H}-\overset{\displaystyle \text{H}}{\underset{\displaystyle \text{H}}{\text{N}}}\!:
$$

$$
\text{H}\cdot + \cdot\ddot{\underset{\cdot\cdot}{\text{Cl}}}\!: \longrightarrow \text{H}-\ddot{\underset{\cdot\cdot}{\text{Cl}}}\!:
$$

1.2. Nuclear Charge

Atoms in molecules often bear a charge, and it is the purpose of this section to describe a method for assigning such charges. Consider the two species formed by proton transfer between a pair of water molecules:

$$
\text{H}-\ddot{\underset{\cdot\cdot}{\text{O}}}-\text{H} + \text{H}-\ddot{\underset{\cdot\cdot}{\text{O}}}-\text{H} \longrightarrow \text{H}-\ddot{\underset{\cdot\cdot}{\text{O}}}\!:^{-} + \text{H}-\overset{\displaystyle \text{H}}{\underset{\cdot\cdot}{\text{O}}}{}^{+}\!-\text{H}
$$

The formal charge on the oxygen of water and its two ionic derivatives can be calculated by using the following formula:

$$
\text{charge} = A - (B/2 + C)
$$

where A is the number of electrons in the outer shell of the neutral atom, B is the number of shared electrons, and C is the number of unshared electrons. The values for A, B, and C are shown below:

$$
\begin{array}{ccc}
\text{H--}\overset{..}{\underset{..}{\text{O}}}\text{--H} & \text{H--}\overset{..}{\underset{..}{\text{O}}}\text{:} & \text{H--}\overset{\overset{\text{H}}{|}}{\underset{..}{\text{O}}}\text{--H}
\end{array}
$$

$A = 6$	$A = 6$	$A = 6$
$B = 4$	$B = 2$	$B = 6$
$C = 4$	$C = 6$	$C = 2$
charge = 0	charge = -1	charge = $+1$

Carbon (in which $A = 4$) can exist in both cationic and anionic forms:

$$
\begin{array}{cc}
\overset{\text{H}\diagdown\quad\diagup\text{H}}{\underset{\underset{\text{H}}{|}}{\text{C}^+}} & \overset{\overset{\text{H}}{|}}{\underset{\underset{\text{H}}{|}}{\text{H--C}{:}^-}}
\end{array}
$$

$A = 4$	$A = 4$
$B = 6$	$B = 6$
$C = 0$	$C = 2$
charge = $+1$	charge = -1

With a little practice, use of the formula becomes unnecessary; one soon is able to assign charge "intuitively."

The common practice of omitting unshared pairs of electrons from organic structures can cause confusion when calculating charge. This is particularly true for carbanions, in which anionic carbon is displayed usually without its unshared pair:

$$CH_3^- \quad \text{implies} \quad {:}CH_3^-$$

Other examples are shown below:

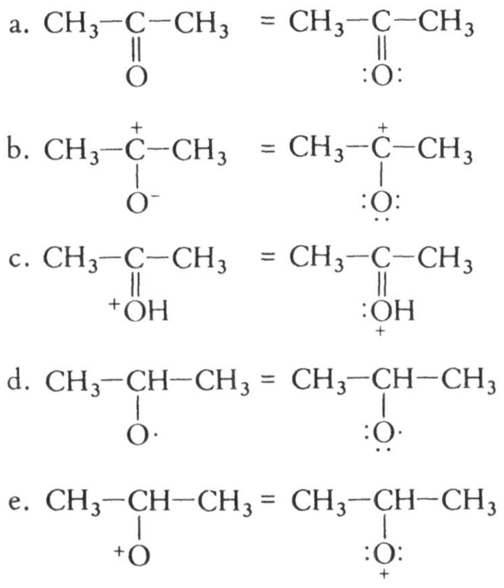

a. $CH_3-\overset{\|}{\underset{O}{C}}-CH_3$ $=$ $CH_3-\overset{\|}{\underset{:\ddot{O}:}{C}}-CH_3$

b. $CH_3-\overset{+}{\underset{O^-}{C}}-CH_3$ $=$ $CH_3-\overset{+}{\underset{:\ddot{O}:}{C}}-CH_3$

c. $CH_3-\overset{\|}{\underset{+\ddot{O}H}{C}}-CH_3$ $=$ $CH_3-\overset{\|}{\underset{:\overset{+}{\ddot{O}}H}{C}}-CH_3$

d. $CH_3-\overset{}{\underset{\overset{|}{O\cdot}}{CH}}-CH_3$ $=$ $CH_3-\overset{}{\underset{\overset{|}{:\ddot{O}\cdot}}{CH}}-CH_3$

e. $CH_3-\overset{}{\underset{\overset{|}{+O}}{CH}}-CH_3$ $=$ $CH_3-\overset{}{\underset{\overset{|}{:\overset{+}{\ddot{O}}:}}{CH}}-CH_3$

Convince yourself, by means of the formula, that oxygen bears a positive charge in either of *two* situations: (1) three covalent bonds and one unshared pair [example (c)] ; (2) one covalent bond and two unshared pairs [example (e)]. Oxygen of the latter type is particularly unstable because it lacks a full octet.

Since electrons can be neither created nor destroyed in any chemical reaction, the charges must balance on both sides of an equation. This fact can assist in the assigning of charge to individual species. For example, species C in the following reaction must be neutral in order to equal the sum of the charges of the reactants:

$$A^+ + B^- \longrightarrow C$$

If the charges in an equation do not balance, this means that either charges have been misassigned or else a charged species has been omitted from the equation. For example, a proton is often not included in the ionization of a carboxylic acid:

$$RCOOH \longrightarrow RCOO^-$$

Charge balance would have been achieved if the equation had been presented in its complete form:

$$RCOOH \longrightarrow RCOO^- + H^+$$

Table 1 lists the charges on oxygen, nitrogen, and carbon in various states.

TABLE 1. Charge on Atoms

Atom	Number of pairs of unshared electrons	Number of groups bonded to atom[a]	Charge	Example
O	1	3	+	$(CH_3)_3\overset{+}{\underset{\cdot\cdot}{O}}:$
	2	2	0	$CH_3\ddot{O}H$
	3	1	–	$CH_3\ddot{O}:^-$
N	0	4	+	$(CH_3)_4N^+$
	1	3	0	$(CH_3)_3N:$
	2	2	–	$(CH_3)_2\ddot{N}:^-$
C	0	4	0	CH_4
	1	3	–	$\overset{-}{:}CH_3$ [b]
	0	3	+	$\overset{+}{C}H_3$
	1	2	0	$:CH_2$

[a] Count double and triple bonds as 2 and 3 groups, respectively. For example, the N in $R-C\equiv N^+-R$ has "4" groups.

[b] According to current notation, carbanions are written without the free pair of electrons (i.e., $^{\ominus}CH_3$).

1.3. Inductive Effects

Although two different atoms may share a pair of electrons to form a covalent bond, they do not necessarily share the electrons equally. Sometimes the electrons will be closer to one of the atoms than the other, and the bond is said to be polarized. Hydrogen chloride has a polarized bond, with the chlorine receiving the greater share of the electrons. This fact is depicted schematically in a number of ways:

$$\text{H} \longrightarrow \text{Cl} \quad\quad \text{H} \ :\!\text{Cl} \quad\quad \text{H}^{+\delta}\!-\!\text{Cl}^{-\delta}$$

In the third representation, δ indicates some fractional charge greater than zero but less than one. The magnitude of δ can be determined from the dipole moment μ of hydrogen chloride, the dipole moment being defined as the partial atomic charge multiplied by the distance of the charge separation d:

$$\mu = \delta \times d \times 10^{18}$$

where μ is in Debye units, δ in electrostatic units, and d in centimeters. The hydrogen–chlorine bond distance is 1.28×10^{-8} cm and the charge on an electron is 4.8×10^{-10} esu. Using these numbers, one can calculate $\mu = 6.15$ D, which would apply if the hydrogen and chlorine were fully ionic (+1 and –1, respectively). In actual fact, hydrogen chloride is *not* fully ionic and thus has a dipole moment of only 1.03 D. This means that the charge on the hydrogen is only 1.03/6.15 or +0.17; chlorine has an identical, but negative, charge. When hydrogen chloride gas is dissolved in water, the chlorine retains both electrons of the covalent bond, leaving behind a proton (or, more properly, a hydronium ion, H_3O^+):

$$\text{H}^{-\delta}\!-\!\text{Cl}^{-\delta} \xrightarrow{\ H_2O\ } \text{H}^+ + \text{Cl}^-$$

The above reaction constitutes our first "mechanism"; electron movement by the electron *pair* is indicated by an arrow.

The ability of an atom to attract electrons is termed *electronegativity*. Thus, chlorine is more electronegative (electron-attracting) than hydrogen. The following sequence gives the relative electronegativities of atoms commonly found in organic chemistry:

$$F > O > N \approx Cl > Br > C \approx H$$

Since oxygen is more electronegative than chlorine, one would expect the oxygen–chlorine bond to be polarized as shown below:

$$O^{-\delta}\!\!-Cl^{+\delta}$$

Carbon, a low-electronegativity atom, displays small differences depending on its hybridization:

$$sp > sp^2 > sp^3$$

The relatively high electronegativity (for a carbon) of the sp-hybridized atom has been explained by its 50% s character (compared to 33% for sp^2 and 25% for sp^3). Since the s orbital is smaller than a p orbital, the sp orbital is smaller (closer to the positively charged nucleus) than an sp^2 or sp^3 orbital. Hence, the sp carbon is the most electronegative.

When the properties of a compound are affected by electronegativity differences, the compound is said to be subject to an *inductive* effect. For example, the dipole moment of hydrogen chloride is the result of an inductive effect by the chlorine. Inductive effects arising from electronegativity differences are extremely important in rationalizing the behavior of organic compounds, as illustrated by the following examples:

(a) The boiling points of four saturated hydrocarbons are given below:

Hydrocarbon	MW	BP
C_5H_{12}	72	36
C_6H_{14}	86	69
C_7H_{16}	100	98
C_8H_{18}	114	126

Boiling points rise steadily as the molecular weight of the hydrocarbon increases. Compare the above data with the boiling points for three oxygen-containing compounds:

Compound	MW	BP
$C_2H_5OC_2H_5$	74	35
C_2H_5OH	46	78
HOH	18	100

There is an *inverse* relationship between molecular weight and boiling point. Water, with a molecular weight of only 18, has a boiling point 65° higher than that of diethyl ether, a much larger molecule. This anomalous boiling point reflects electrostatic attraction among the polarized water molecules:

$$
\begin{array}{c}
\quad\;\; \text{H} \\
\quad\;\; | \\
\text{H---O--H---O--H} \\
\quad | \qquad\qquad\; | \\
\text{H--O} \qquad\quad\; \text{H}
\end{array}
$$

Ether is also subject to polarization, but the carbons, being somewhat shielded by their protons, do not associate with negatively

$$CH_3CH_2 \longrightarrow O \longleftarrow CH_2CH_3$$

charged centers as effectively as do the more exposed protons of water.

(b) The S_N2 displacement reaction between hydroxide ion and methyl bromide can be rationalized by inductive effects. Since

$$HO^- \longrightarrow \overset{+\delta}{C}H_3 \overset{-\delta}{-}Br \longrightarrow HO-CH_3 + Br^-$$

bromine is more electronegative than carbon, the latter bears a partial positive charge. Hydroxide ion, an anionic nucleophile, attacks the oppositely charged carbon and displaces bromide. Bromine departs with both electrons of the original C—Br bond.

(c) Carbonium ions can be formed by protonation of a double bond:

$$R-CH \overset{H^+}{=\!\!=} CH_2 \longrightarrow R-\overset{+}{C}H$$

The π electrons of the double bond are used to form a new C—H σ bond. It is found that the rate of formation of the carbonium ion intermediate is much faster when R = CH$_3$ than when R = CF$_3$:

$$\begin{matrix} & H & & & F & \\ & | & & & | & \\ H- & C & -CH=CH_2 & > & F-\overset{+\delta}{C}-CH=CH_2 \\ & | & & & | & \\ & H & & & F & \end{matrix}$$

The three highly electronegative atoms create a positive charge on the carbon to which they are bound. Formation of the carbonium ion from $CF_3CH=CH_2$ is thereby inhibited because two atoms of similar charge do not "like" being side by side:

$$\overset{+\delta}{F_3C}-\overset{+}{C}H-CH_3 \quad \text{(unfavorable)}$$

(d) Chloroacetic acid is a stronger acid than acetic acid:

$$CH_3COOH \quad pK_a = 4.76$$

$$ClCH_2COOH \quad pK_a = 2.86$$

Electron withdrawal by the chlorine shifts electron density away from the acidic proton, thereby favoring proton loss from the acid:

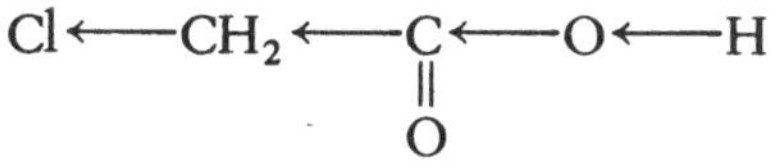

$$Cl \longleftarrow CH_2 \longleftarrow \underset{\underset{O}{\|}}{C} \longleftarrow O \longleftarrow H$$

As one might expect, the further away the chlorine is from the reactive site, the smaller the inductive effect. Thus, chloropropionic acid has a pK_a close to that of acetic acid:

$$ClCH_2CH_2COOH \quad pK_a = 4.08$$

(e) Carbonium ion stability in solution obeys the order tertiary > secondary > primary:

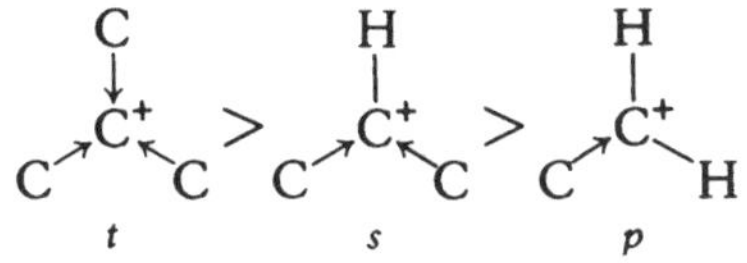

Alkyl groups seem to donate electron density relative to a proton, and therefore the tertiary carbonium ion with three such groups is the most stable. Although this is certainly not the complete story, the "electropositive" nature of alkyl groups relative to a proton has been widely used to rationalize and predict organic behavior.

1.4. Resonance Effects

We will devote considerable attention to resonance for two reasons. First, resonance is one of the most important effects in organic chemistry. The relative reactivity of a series of compounds can often be rapidly and accurately assessed by simple resonance considerations. Second, the drawing of resonance contributors provides valuable practice in the "pushing" of electrons, a skill necessary—as we shall see—in developing reaction mechanisms.

The concept of resonance is easiest to introduce by means of an example, benzene. Resonance theory says that benzene can be represented by two *contributors* which are interconvertible merely by shifting the mobile π electrons:

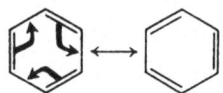

Benzene is *not* a mixture of these contributors but rather a "hybrid" of the two. That is to say, the electronic structure of benzene lies intermediate between the fictitious, but drawable, contributors. Thus, the carbon–carbon double bonds in benzene are neither single nor double but somewhere in between. Since it is difficult to depict structures with fractional bond orders, one does the next best thing: nonexistent contributors are drawn which resemble, at least to some extent, the actual molecule. Resonance contributors are always indicated by a double arrow ($\longleftrightarrow$). This should not be confused with the symbol for an equilibrium ($\rightleftharpoons$) relating species with a discrete physical existence.

All resonance hybrids are more stable than the contributors would be if the contributors actually existed. For example, benzene is more stable (less energetic) than "cyclohexatriene" (an imaginary compound in which the π electrons are not delocalized). The stability associated with delocalization of electrons in a π

system is called the *resonance energy*. Benzene, with a particularly large resonance stability, reacts much more slowly with bromine than does the localized double bond of ethylene.

Benzene is characterized by two equivalent resonance forms which contribute equally to the overall benzene structure (i.e., benzene resembles one no more than the other). But resonance forms need not contribute equally. For example, the resonance contributors for the enolate anion are not identical:

$$\overset{\displaystyle \overset{O^-}{|}}{RC}=CH_2 \longleftrightarrow \overset{\displaystyle \overset{O}{\|}}{RC}-CH_2^-$$

Undoubtedly, the enolate anion resembles more closely the contributor with the negative charge on the electronegative oxygen.

The following rules should help the student in drawing resonance contributors and in assessing their relative importance.

Rule 1. There can be no shift in the position of atoms from one resonance contributor to another. Bond angles remain constant. Only electron density changes.

resonance contributors

different compounds

different compounds

Rule 2. Keep in mind the electronegativities of atoms in assessing the importance of contributors.

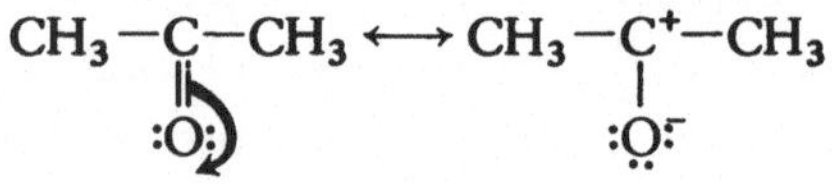

$$CH_3-\overset{\underset{\displaystyle :\overset{\cdot\cdot}{O}:}{\|}}{C}-CH_3 \longleftrightarrow CH_3-\overset{\underset{\displaystyle :\overset{\cdot\cdot}{O}:^{-}}{|}}{C^{+}}-CH_3$$

some importance

$$CH_3-\overset{\underset{\displaystyle :\overset{\cdot\cdot}{O}:}{\|}}{C}-CH_3 \longleftrightarrow\!\!\!\!\!/ \, CH_3-\overset{\underset{\displaystyle :\overset{\cdot\cdot}{O}:^{+}}{|}}{\bar{C}}-CH_3$$

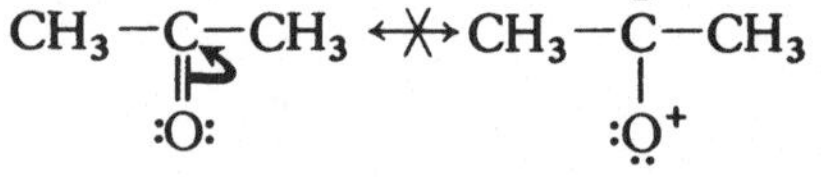

no importance

The contributor of acetone, in which the electronegative oxygen has a positive charge and (worse) only six electrons in its outer shell, is of no importance. The only reasonable dipolar structure has a negative oxygen and positive carbon; this contributor predicts that the central carbon of acetone should possess a partial positive charge (inductive effect arguments lead to the same conclusion). Thus it is not surprising that acetone is attacked at the electron-deficient carbon by hydroxide ion and other electron-rich nucleophiles:

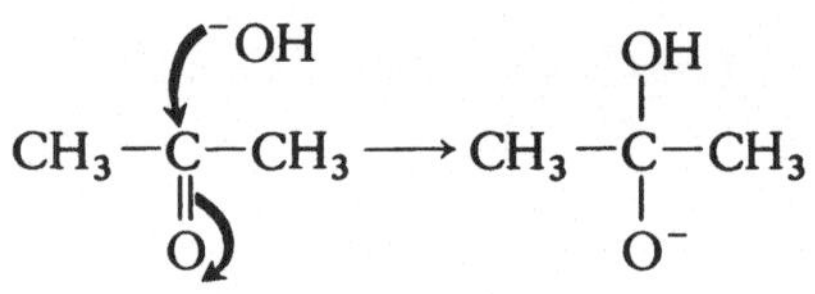

$$CH_3-\overset{\underset{\displaystyle O}{\|}}{C}-CH_3 \longrightarrow CH_3-\overset{\underset{\displaystyle O^{-}}{|}}{\overset{\overset{\displaystyle OH}{|}}{C}}-CH_3$$

Another example of an unlikely resonance contributor is shown below:

$$H_2C\!\!=\!\!CH_2 \longleftrightarrow H_2\overset{+}{C}-\bar{C}H_2$$

Since the two carbons have identical electronegativities, there is little reason for expecting one of them to accept both electrons of the π bond leaving the other carbon electron deficient.

Rule 3. No first-row atom, including carbon, oxygen, and nitrogen, may contain more than eight electrons in its outer shell.

Since oxygen is more electronegative than carbon, it might be tempting to move electrons in the direction of oxygen:

impossible structure

However, this leads to an impossible structure with an oxygen having 10 outer electrons. The error is best avoided by including all unshared pairs of electrons of the contributors; in this way a violation of the octet rule becomes more apparent. Note that although the phenol oxygen cannot accept additional electrons, it can donate a pair to the ring:

The dipolar resonance contributors are reasonable because the oxygen bears a positive charge by virtue of sharing a pair of electrons (as it does with a proton in hydronium ion). Furthermore, the oxygen octet is maintained. This is in contrast to the unstable type of positively charged oxygen which has only six outer electrons (illustrated in Rule 2).

Rule 4. Other factors being equal, there is more resonance stabilization when charge is not created.

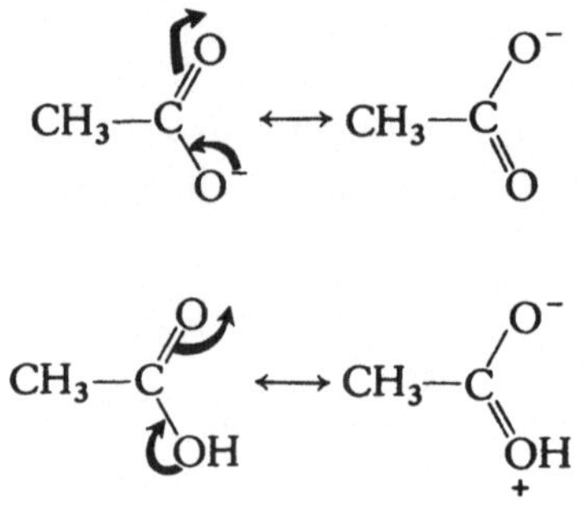

Acetate ion is highly resonance stabilized. Since the two anionic structures contribute equally, each oxygen of the hybrid bears exactly one-half a negative charge. Acetic acid is also resonance stabilized but to a much lesser extent because of charge creation; it takes energy to create and separate two opposite charges. Along similar lines, one would reject as absurd the third resonance contributor shown below:

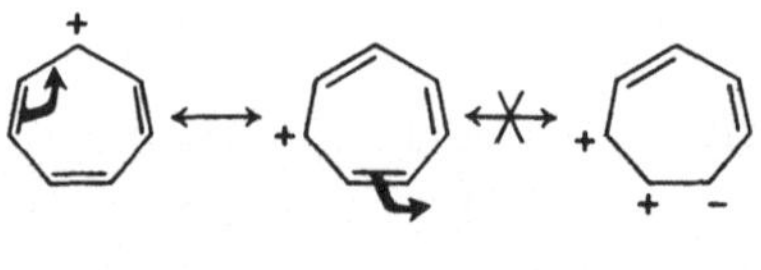

Likewise the third form would contribute nothing to the overall structure of N-methylpyridinium ions:

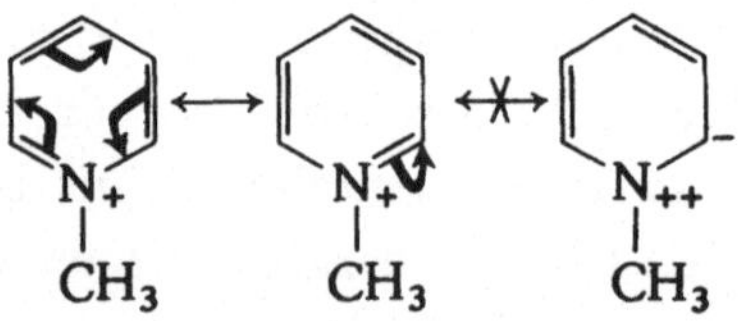

Rule 5. Stabilization is greatest when there are two or more

equivalent structures of lowest energy.

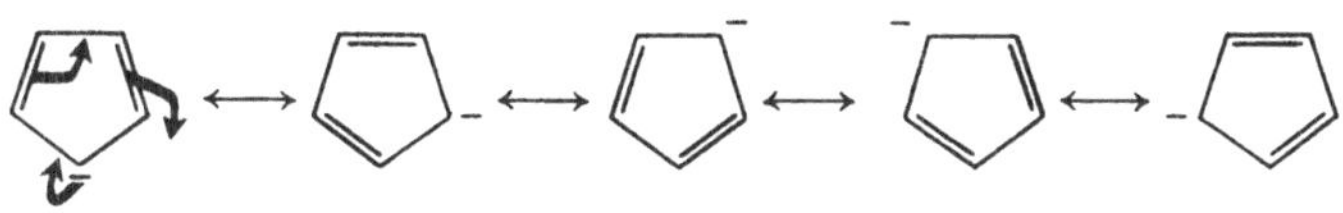

Cyclopentadiene is about 10^{30} times more acidic than cyclopentane because the anion which forms upon loss of a proton has five identical contributors. Each carbon must bear only one-fifth of a negative charge; distribution of the negative charge among several atoms stabilizes the species.

Important equivalent resonance contributors of benzene and of acetate have already been mentioned; another example is given below:

$$R_2\ddot{N}-CH=CH-CH=\underset{+}{N}R_2 \longleftrightarrow R_2\overset{+}{N}=CH-CH=CH-\ddot{N}R_2$$

Cyclobutadiene illustrates an interesting exception to the equivalency rule. This species is extremely unstable despite the fact that one may draw two identical resonance contributors:

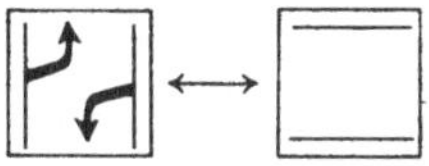

It turns out that a monocyclic conjugated system must possess $4n + 2$ π resonating electrons ($n = 0, 1, 2$, etc.) for appreciable aromatic-type stabilization (Hückel's rule). Thus, cyclopentadienyl cation (unlike the corresponding anion described above) is not well stabilized by resonance since $4 \neq 4n + 2$:

 etc.

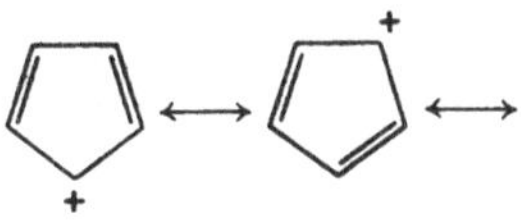

Rule 6. Important resonance contributors must have reasonable bond lengths and angles.

By merely "pushing electrons" one can arrive at a structure which is formally a resonance contributor of benzene:

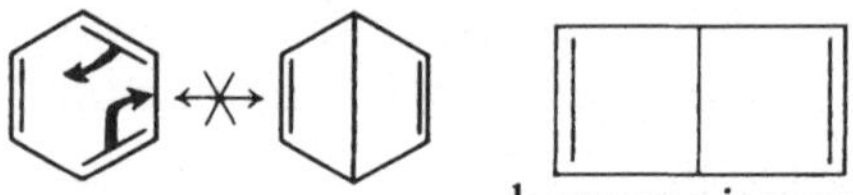

benzene isomer

It is, however, absolutely unimportant because the length of the central single bond is unrealistic (as can be seen by comparing it with the other bonds in the structure). The bicyclic compound shown above is an *isomer* of benzene. One could not write this structure as a benzene contributor because of Rule 1. The atomic geometry of all contributors must be the same, and this clearly is not the case for hexagonal benzene and its rectangular isomer.

The following sets of structures will further illustrate the use of electron pushing to devise resonance contributors:

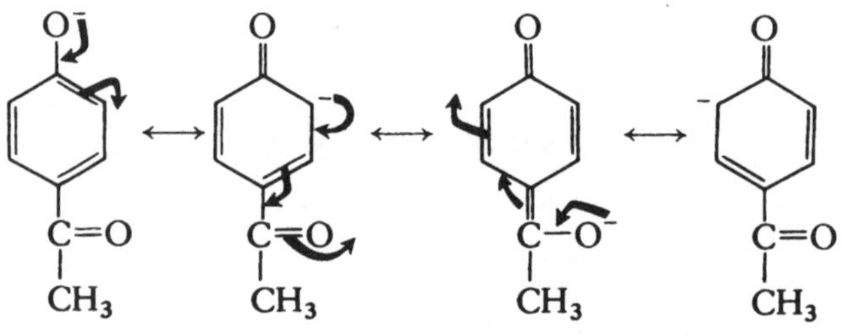

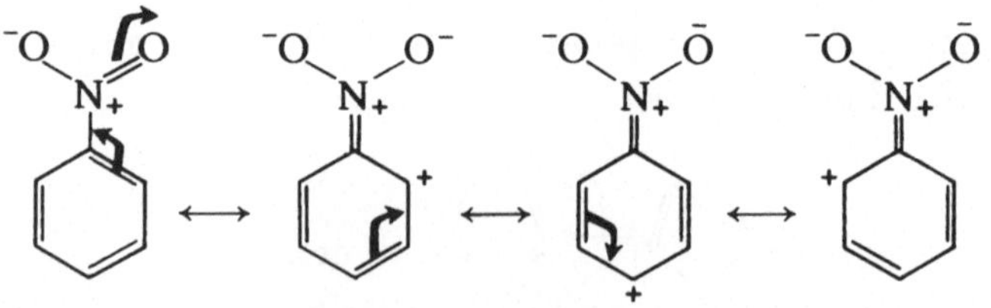

$$R-\overset{\curvearrowleft}{\underset{\underset{O^{\curvearrowright}}{\|}}{C}}-\overset{}{N}R_2 \longleftrightarrow R-\underset{\underset{O^-}{|}}{C}=\overset{+}{N}R_2$$

$$R-\underset{\underset{O}{\|}}{C}\curvearrowright CH\overset{}{=}CH\overset{\curvearrowleft}{\overset{}{N}}R_2 \longleftrightarrow R-\underset{\underset{O^-}{|}}{C}=CH-CH=\overset{+}{N}R_2$$

1.5. Acidity and Basicity

The acidity of a substance is reflected by the position of the following equilibrium:

$$HA \rightleftharpoons H^+ + A^-$$

An equilibrium constant K relates the concentrations (activities) of the three species:

$$K = \frac{[H^+][A^-]}{[HA]}$$

Clearly, the stronger the acid the larger the value of K. Acetic acid, a weak organic acid, is characterized by a K of only 1.8×10^{-5} M. Since K values of organic acids are typically very small, it is more convenient to describe acidity in terms of pK_a:

$$pK_a = -\log K$$

Thus, the pK_a of acetic acid is 4.76. Because of the negative sign in the pK_a definition, the *stronger* the acid the *smaller* the pK_a. For example, formic acid, with a pK_a of 3.75, is a stronger acid than acetic acid by 1 pK_a unit (a factor of 10 in K). One can easily derive the following relationship:

$$pK_a = pH - \log \frac{[A^-]}{[HA]}$$

The above equation is useful because it points out a convenient fact: when pH = pK_a, there exists 50% HA and 50% A^-. For example, a 1 M solution of acetic acid adjusted to pH 4.76 will contain 0.5 M HOAc and 0.5 M AcO^-. Raising the pH to 5.76 and 6.76 increases the proportion of AcO^- to 90% and 99%, respectively. Decreasing the pH to 3.76 and 2.76 increases the concentration of the acid form, HOAc, to 90% and 99%, respectively. At pH 14, the amount of free acid would be too small to measure.

Since many mechanisms that we will consider later depend on acid–base properties, it is important to develop a feeling for the subject. Both inductive and resonance effects play a role in determining the pK_a of organic compounds. Consider the decreasing order of acidity in the following sequence:

Acidity: CH_3COOH > > CH_3OH

pK_a: 4.8 10 15

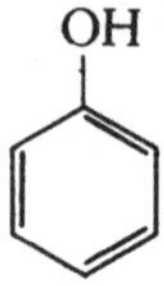

Methanol is an extremely weak acid because the negative charge in the conjugate base (methoxide ion) is forced to reside on a single atom. With phenolate, the charge can be distributed by resonance:

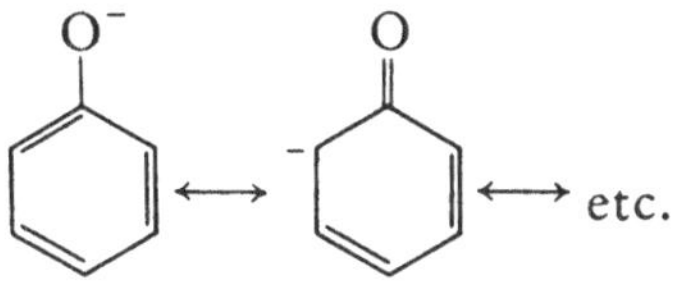

This resonance stabilization serves to drive the ionization equilibrium to the right, thus explaining why phenol is 5 pK_a units more acidic than methanol. Of course, phenol in the protonated state is also resonance stabilized:

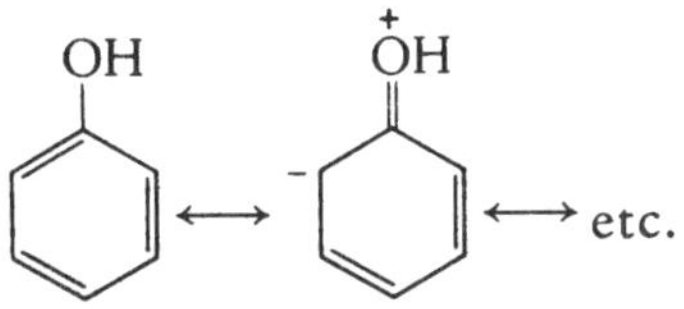

However, this resonance (which creates charge) is less effective than that in phenolate (see Rules 1-6 on resonance). Acetate anion is stabilized by a particularly favorable resonance in which the two oxygens share equally the negative charge; proton loss from acetic acid to form acetate therefore occurs more readily than proton loss from phenol. Note that a mixture of acetic acid, phenol, and methanol dissolved in a buffer of pH 7.0 would contain mainly acetate anions and the conjugate acids of phenol and methanol.

A *para* acetyl group on the phenol ring lowers the pK_a (increases the acidity) to 8.0. This can be rationalized by the additional resonance stabilization of the phenolate anion resulting from the acetyl group:

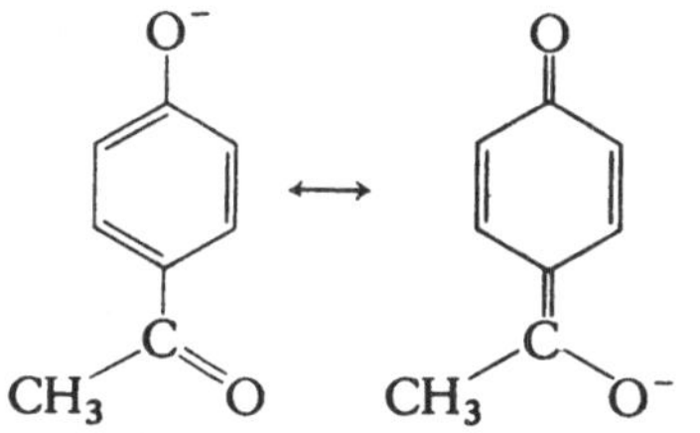

So-called "carbon acids" play a major role in synthetic organic chemistry and in many of the mechanisms that we will discuss. Several examples are given below with the acidic proton underlined:

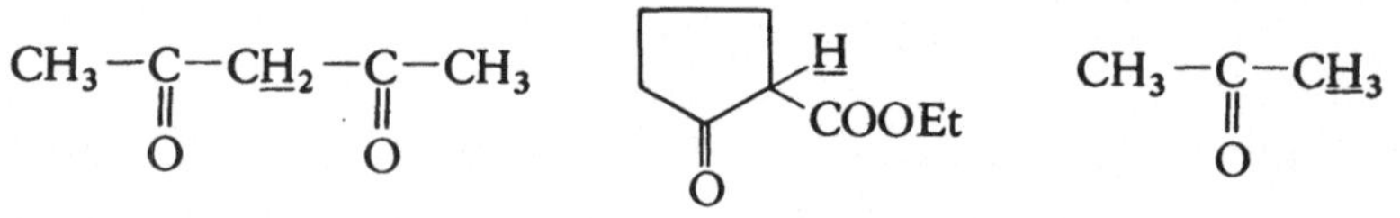

$$CH_3-\underset{\underset{O}{\|}}{C}-C\underline{H}_2-\underset{\underset{O}{\|}}{C}-CH_3$$

$$CH_3-\underset{\underset{O}{\|}}{C}-C\underline{H}_3$$

pK_a: 9   10   20

Acetylacetone has a pK_a even lower than that of phenol because of the following resonance stabilization of the enolate anion:

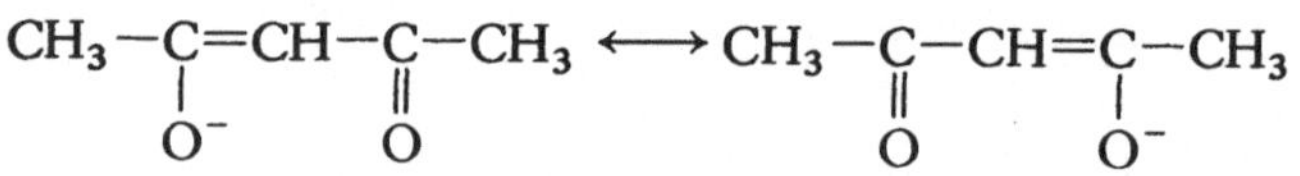

$$CH_3-\underset{\underset{O^-}{|}}{C}=CH-\underset{\underset{O}{\|}}{C}-CH_3 \longleftrightarrow CH_3-\underset{\underset{O}{\|}}{C}-CH=\underset{\underset{O^-}{|}}{C}-CH_3$$

Acetone, with only one carbonyl to accept the negative charge, has a pK_a 11 units higher than that of acetylacetone. This means that in water (with a practical pH limit of 14) one can never achieve more than a tiny fraction of acetone in the enolate state. Nonetheless the acetone enolate can be chemically important since it is far more reactive than acetone itself. For example, acetone is brominated in basic water according to the following scheme:

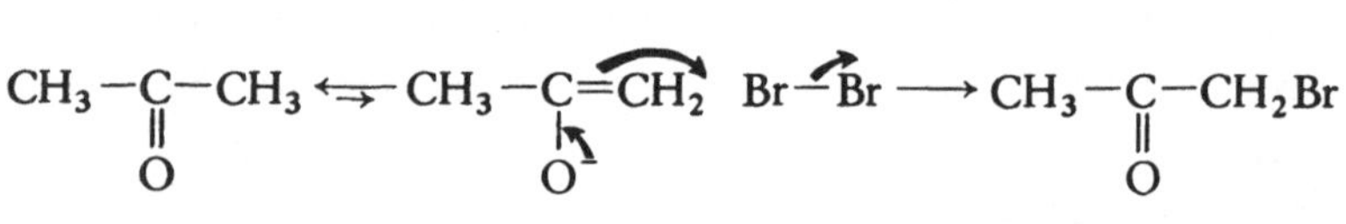

$$CH_3-\underset{\underset{O}{\|}}{C}-CH_3 \rightleftharpoons CH_3-\underset{\underset{O^-}{|}}{C}=CH_2 \quad Br-Br \longrightarrow CH_3-\underset{\underset{O}{\|}}{C}-CH_2Br$$

Although the preequilibrium ionization is unfavorable in water, the small amount of enolate that does form reacts quickly with bromine. Simultaneously, more enolate is created to reestablish the equilibrium concentrations, and ultimately all the acetone can be brominated if enough bromine is present.

Use of a base stronger than hydroxide ion gives higher concentrations of enolate. For example, t-butoxide ion (prepared from t-butanol and potassium metal) effectively abstracts protons from acetone and thereby accelerates the formation of enolate-derived products:

$$CH_3-\underset{\underset{O}{\|}}{C}-CH_2 \quad H \quad O-\underset{\underset{CH_3}{|}}{\overset{\overset{CH_3}{|}}{C}}-CH_3 \longrightarrow CH_3-\underset{\underset{O^-}{|}}{C}=CH_2$$

t-Butoxide ion is a stronger base than hydroxide ion because the methyl groups (a) direct electron density toward the oxygen and (b) sterically impede the solvation (stabilization) of the oxygen anion. Amide ion, NH_2^-, and butyl lithium, $CH_3CH_2CH_2\bar{C}H_2Li^+$, are two other extremely powerful bases which are commonly employed with weak carbon acids.

In assessing relative acidity and basicity, it is helpful to realize that the stronger the conjugate acid, the weaker the conjugate base (and vice versa). Since acetic acid is a stronger acid than methanol, methoxide ion is a stronger base than acetate ion. Nowadays, one customarily reports the basicity of amines in terms of the acidity of the protonated form. Although this may seem confusing, it permits the use of only a single parameter (pK_a) rather than two (pK_a and pK_b). A single example will suffice to illustrate the point:

$$\begin{array}{cc} NH_3^+ & NH_3^+ \\ \hexagon & \hexagon \\ & NO_2 \end{array}$$

pK_a: 5.1   1.0

The pK_a refers to the following equilibrium:

$$RNH_3^+ \rightleftharpoons RNH_2 + H^+$$

Anilinium ion is a *weaker* acid than *p*-nitroanilinium ion; therefore aniline is a *stronger* base than *p*-nitroaniline. This is exactly what one would expect since the latter has a nitro group which withdraws the crucial electron pair on the amino group into the aromatic system:

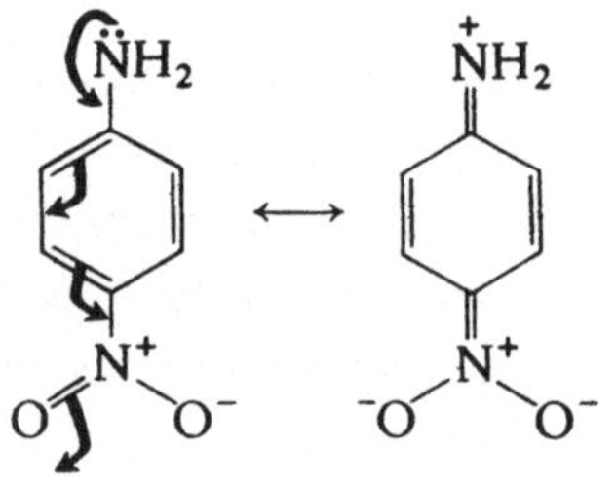

Any compound with a neutral oxygen atom possesses a basic center by virtue of the unshared pairs of electrons on the oxygen:

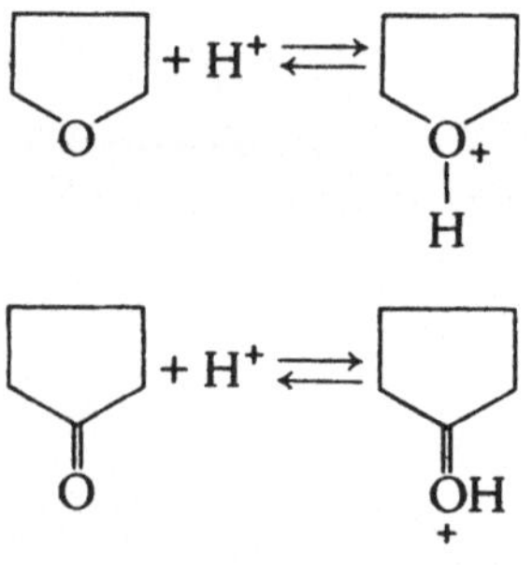

In general, alcohols, ethers, ketones, and other neutral-oxygen species are very weak bases, and even in strong acid (e.g., 1 N HCl) only a small percentage of the molecules are protonated as shown above. In some of the mechanisms discussed later on we will see how these protonated-oxygen bases can be chemically important despite their low concentrations.

Thus far we have viewed an acid as a proton donor and a base as a proton acceptor. But according to the more general "Lewis" definition, an acid can be defined as any species which accepts a pair of electrons. A proton fits this definition, of course, because it can coordinate with, for example, an unshared pair of electrons of water:

$$H-\overset{..}{\underset{\underset{H}{|}}{O}}: + H^+ \longrightarrow H-\overset{+}{\underset{\underset{H}{|}}{O}}-H$$

Lewis acids also include atoms with unfilled valence shells, such as Al in $AlCl_3$ or B in BF_3 (which have only six outer electrons):

$$F^- \rightharpoonup \overset{\underset{\textstyle F}{|}}{\underset{\underset{\textstyle F}{|}}{B}}-F \longrightarrow F-\overset{\underset{\textstyle F}{|}}{\underset{\underset{\textstyle F}{|}}{B^-}}-F$$

$$HO^- + AlCl_3 \longrightarrow HOAlCl_3^-$$

Such nonprotonic acids are mentioned here because they are involved in many organic reactions.

1.6. Oxidation–Reduction

When faced with a new organic reaction, one must first determine whether an oxidation–reduction has occurred. If an organic compound has been oxidized, then obviously some other compound or reagent has been reduced (and vice versa). Clearly, inspection of the oxidation states of all reactants and products is essential to the understanding of a reaction. Oxidation of almost all organic compounds involves either the gaining of oxygen, nitrogen, chlorine, etc., or the loss of hydrogen atoms. Several examples illustrate this point:

Oxidations:

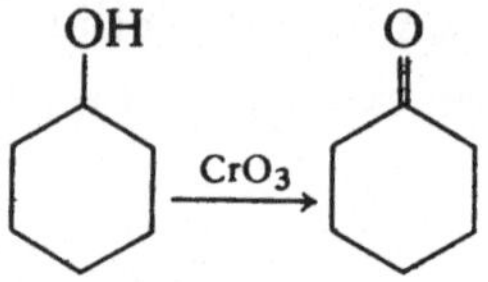

$$CH_3-CH_3 \xrightarrow[h\nu]{Cl_2} CH_3-CH_2-Cl$$

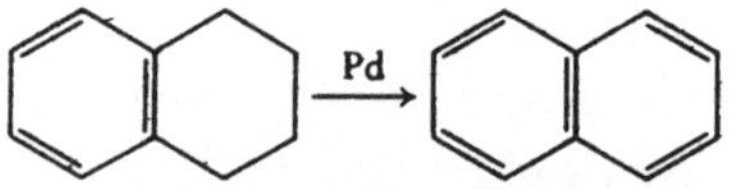

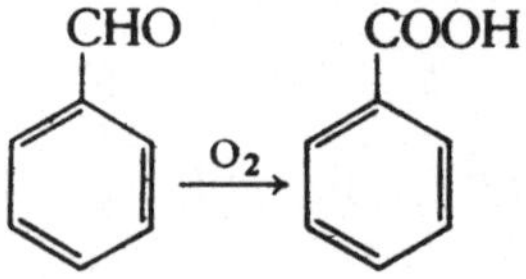

Similarly, reductions entail the gaining of hydrogen and, in many cases, the loss of a heteroatom such as oxygen or chlorine.

Reductions:

$$CH_2{=}CH-CH_3 \xrightarrow{H_2/Pt} CH_3-CH_2-CH_3$$

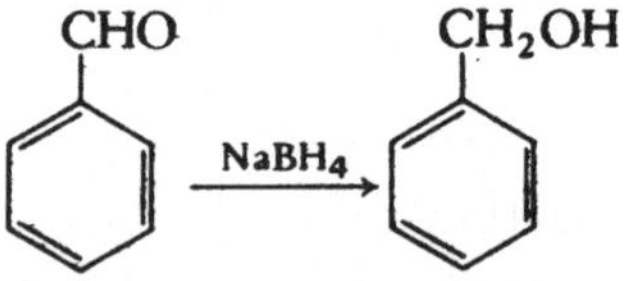

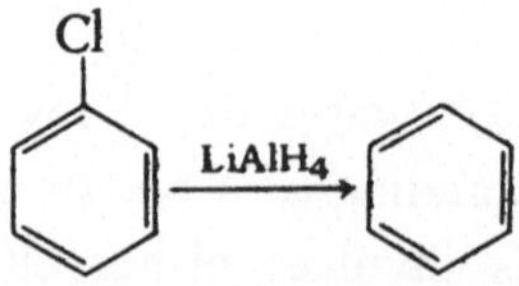

Note that dehydration of an alcohol to an olefin is *not* an oxidation–reduction process:

This is apparent from the fact that no reagent is present which is oxidized or reduced in the course of the reaction. In a sense, one can consider a dehydration an internal oxidation–reduction (one carbon being oxidized and the other reduced); the net effect is no change in the oxidation state of the molecule.

The following procedure can be followed if there is doubt as to the oxidation state of a molecule: (1) Imagine a water molecule being added onto all unsaturated areas in the compound. If the compound has a ring, imagine that the ring is broken at any point with the addition of water. (2) Count the number of heteroatoms in the product of the imaginary water additions, including the nitrogens, chlorines, etc., originally present. The larger the number of heteroatoms, the more highly oxidized the molecule. The oxidation numbers (given in italics below the structures) of several compounds are:

The oxidation number of 2 for formaldehyde was arrived at by the water addition process:

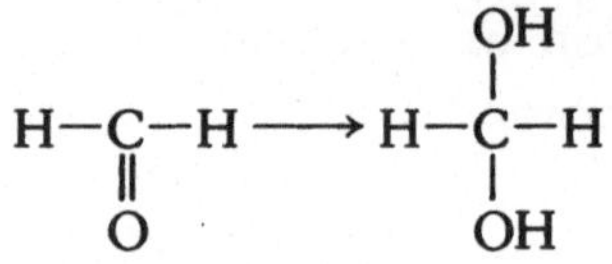

$$
\underset{\text{O}}{\overset{\text{}}{\text{H}-\text{C}-\text{H}}} \longrightarrow \underset{\text{OH}}{\overset{\text{OH}}{\text{H}-\text{C}-\text{H}}}
$$

In order to convert formaldehyde into methanol (a substance with a *smaller* oxidation number) one would have to use a *reducing* agent. No reducing agent would be required to convert formaldehyde into an acetal:

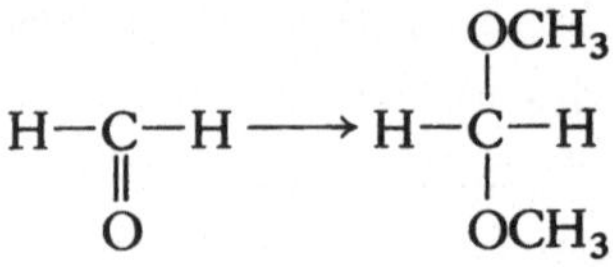

$$
\underset{\underset{2}{\text{O}}}{\overset{}{\text{H}-\text{C}-\text{H}}} \longrightarrow \underset{\underset{2}{\text{OCH}_3}}{\overset{\text{OCH}_3}{\text{H}-\text{C}-\text{H}}}
$$

The ring-opening rule mentioned in connection with calculating oxidation numbers stems from the fact that conversion of, say, hexane to cyclohexane constitutes an oxidation (and the reverse a reduction):

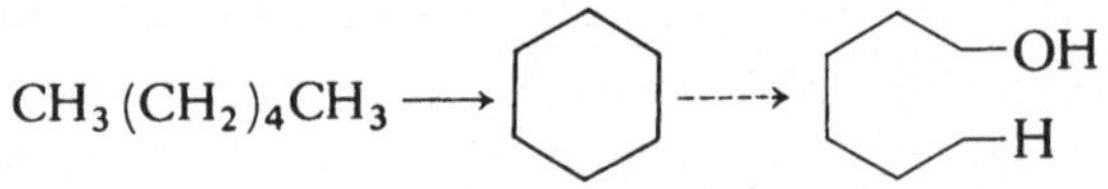

$$
\text{CH}_3(\text{CH}_2)_4\text{CH}_3 \longrightarrow
$$

0   *1*

Three oxidation–reduction reactions are shown below:

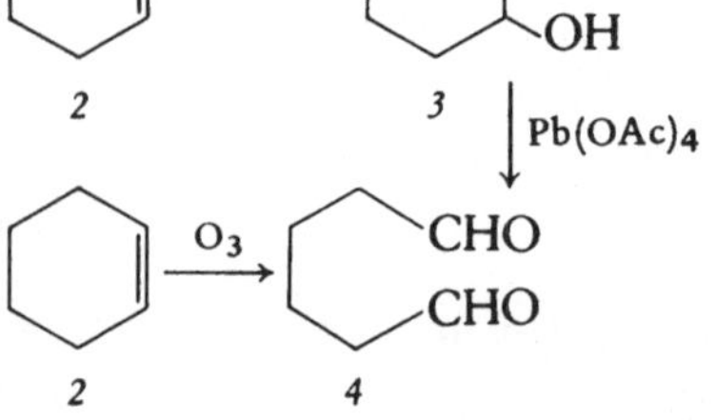

2   *3* H_3O^+

2   *3* $Pb(OAc)_4$

2   *4*

All these reactions (except the hydrolysis of the epoxide to the glycol) involve an increase in oxidation number and hence an oxidative process.

1.7. Reaction Mechanisms

Most organic reactions are not simple one-step, one transition state processes. Often, several intermediates of varying degrees of stability are involved as reactants are converted into products. An organic chemist must be able to guess what these intermediates might be; in other words, he must know how to formulate reasonable mechanisms for complex organic transformations. This skill is necessary for the design of synthetically useful reactions and for the pursuit of physical organic chemistry.

The term "organic reaction mechanism" means different things to different chemists. Physical organic chemists propose detailed mechanisms on the basis of how reactions are affected by changes in concentration, temperature, reactant structure, catalyst, etc. Many reactions in this book are rather complex in nature, and there are little or no such data available. The proposed mechanisms for these reactions can be judged only according to whether or not bond formation and bond cleavage in the various steps are consistent with sound chemical principles. Careful study of this book will, we hope, familiarize the student with these principles and provide a "feel" for organic chemistry.

Fortunately, there are relatively few basic reaction types in organic chemistry. Complex transformations are usually nothing more than a sequence of these basic reactions. The problem, then, becomes one of delineating the sequences. It is the purpose of this book to provide the student with practice in breaking down complicated processes into a series of simple reactions.

In the next section, the student will find problems which can be worked to ensure that the principles covered in this chapter are understood. Answers to all the problems are given at the end of the chapter. In Chapter 2, reaction mechanisms are discussed in detail.

1.8. Problems

1. Determine the number of unshared pairs of electrons on the indicated atoms of the following compounds:

   a. Oxygen: CH_3CHO

   b. Oxygen: $CH_3-\overset{\displaystyle \underset{+OH}{\|}}{C}-CH_3$

   c. Oxygen: $CH_3-\underset{\underset{O^-}{|}}{CH}-CH_3$

   d. Oxygen:

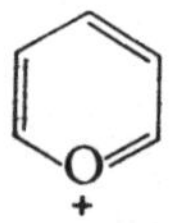

   e. Nitrogen: $CH_3CH_2C\equiv N$

   f. Nitrogen: CH_3NO_2

   g. Nitrogen:

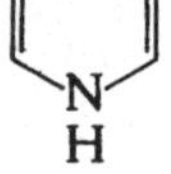

   h. Nitrogen: NH_2^-

   i. Nitrogen: $CH_3-\overset{\displaystyle \underset{N}{\|}}{C}-CH_3$ with N bonded to OH

   j. Carbon: $^-O-\overset{\displaystyle \underset{O}{\|}}{C}-O^-$

k. Carbon (anion): $H-C\equiv C^-$

l. Carbon: CH_3^+

---

2. Are the indicated atoms of the following compounds positive, negative, or neutral? All unshared pairs are shown.

a. Nitrogen:

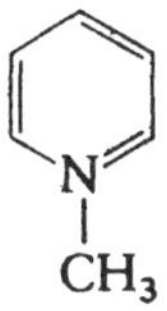

b. Nitrogen: $R_2\ddot{N}:$

c. Nitrogen: $R_2N:$

d. Oxygen: $CH_3-\ddot{N}=\ddot{O}$

e. Oxygen and sulfur: $CH_3-\ddot{S}-CH_3$
                              |
                             $:O:$

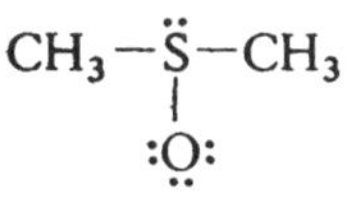

f. Oxygen: $CH_3-\overset{\overset{\displaystyle CH_3}{|}}{\underset{..}{O}}-CH_3$

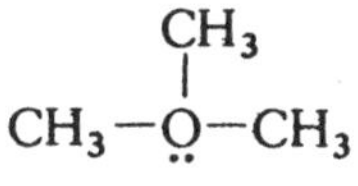

g. Carbon: $H-\ddot{C}-H$

h. Carbon: $R_3C:$

i. Oxygen: $CH_3-\ddot{O}\cdot$

---

3. Given below are several atoms and their atomic numbers. Determine their valences and the number of unshared pairs of electrons in the neutral alkyl derivatives of the atoms.

a. B, A.N. = 5
b. Al, A.N. = 13

    c. F,   A.N. = 9
    d. S,   A.N. = 16
    e. P,   A.N. = 15

---

4. Explain the following differences in pK_a:

a.

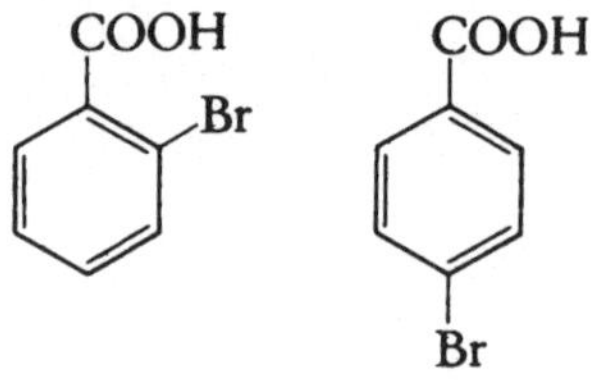

    pK_a: 2.85   4.00

b.

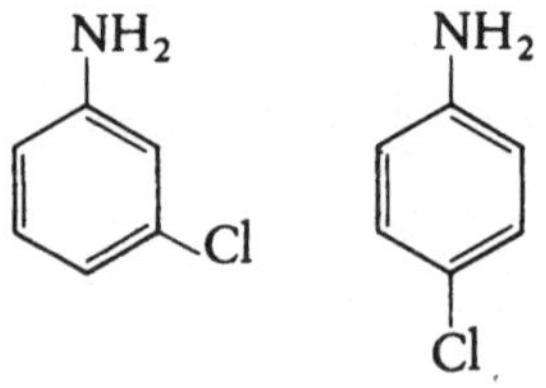

    pK_a: 3.32   3.81

---

5. Show how the following resonance contributors may be interconverted by "electron pushing."

a.

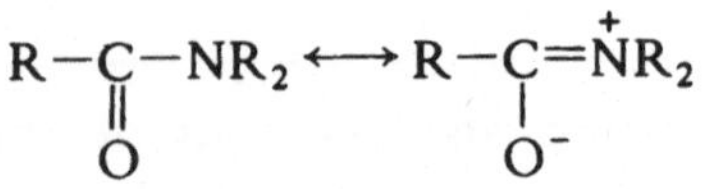

b.

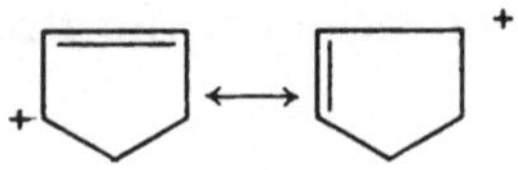

c.

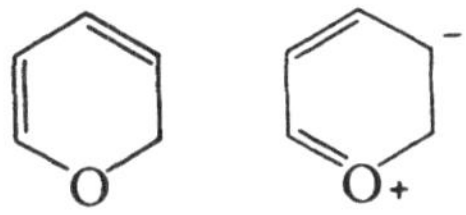

d.

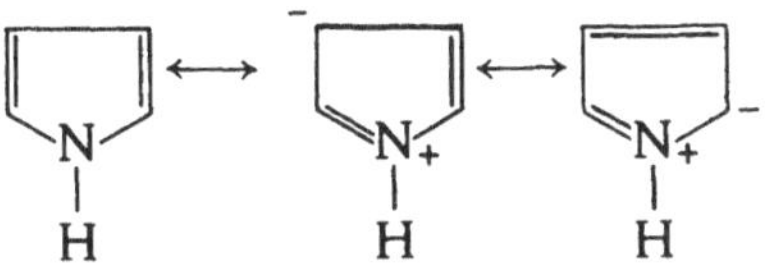

e.

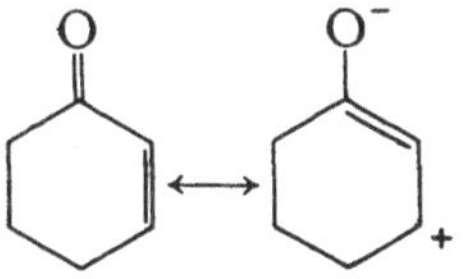

f.

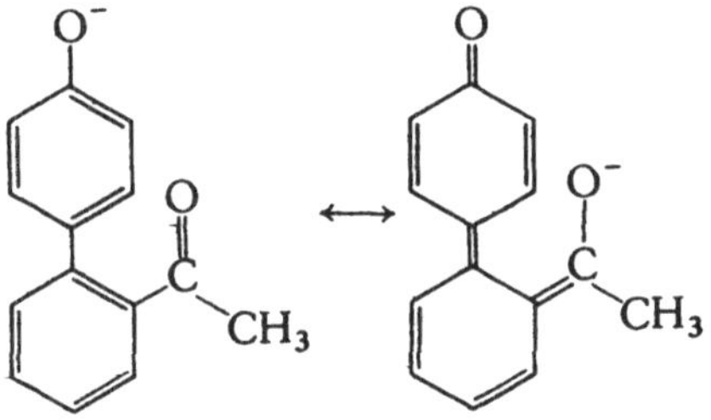

---

6. Draw a reasonable charged resonance contributor for each of the following:

a. $CH_3-NH-C\equiv N$

b. $CH_3-\overset{\displaystyle O}{\underset{\displaystyle \|}{C}}-OCH_3$

c.

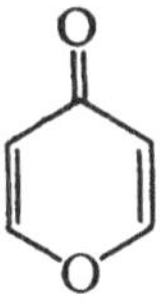

d. 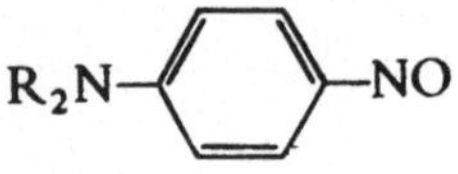 R₂N—⟨benzene ring⟩—NO

e.

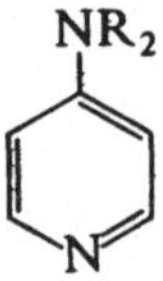

NR₂

f.

HO

O

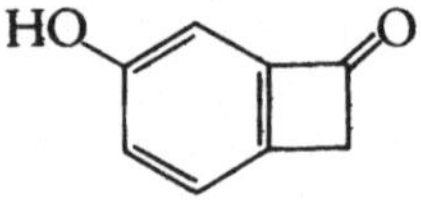

7. Indicate whether the following pairs of structures are resonance contributors, isomers, or the same compound.

a.

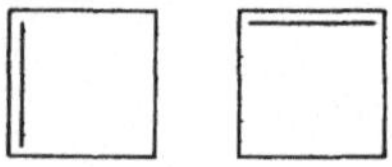

b.

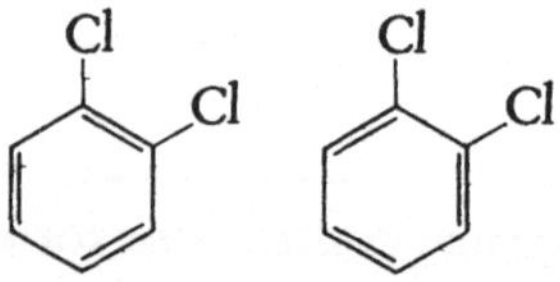

c. CH₂=CH   CH—CH
              \\CH=CH₂   CH₂   \\CH₂

d.

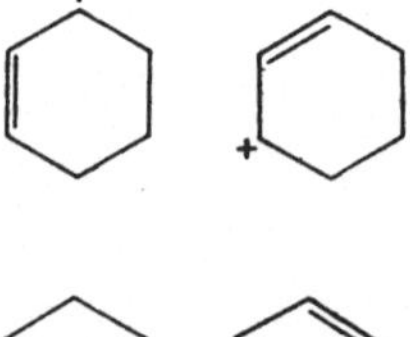

e.

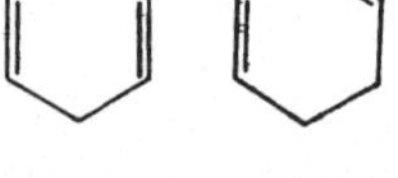

f.

g. $CH_3-\overset{\overset{\displaystyle O}{\|}}{C}-CH_3$ $CH_3-\overset{\overset{\displaystyle}{|}}{\underset{\underset{\displaystyle OH}{|}}{C}}=CH_2$

h.

i. $CH_3CH_2C{\equiv}CH$ $CH_3CH{=}C{=}CH_2$

---

8. Explain why the right-hand members of each of the following are *unreasonable* resonance contributors.

a. $CH_3{\overset{\curvearrowright}{\frown}}CH_3 \longleftrightarrow\!\!\!\!\!\times \overset{+}{C}H_3\ \bar{C}H_3$

b. $CH_3-\overset{\curvearrowright}{\underset{\displaystyle ..}{\ddot{O}}}-CH_3 \longleftrightarrow\!\!\!\!\!\times CH_3-\overset{+}{\ddot{O}}{=}\bar{C}H_3$

c. $\longleftrightarrow\!\!\!\!\!\times$ $\overset{\displaystyle H}{\underset{\displaystyle H}{\overset{\displaystyle |}{\underset{\displaystyle |}{\begin{array}{c}CH_2\ C\\ \|\ \ \ \||| \\ CH_2\ C\end{array}}}}}$

d. $\longleftrightarrow\!\!\!\!\!\times$

e.

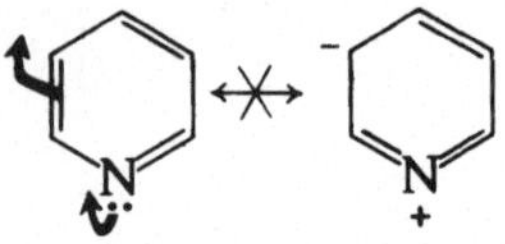

---

9. Draw the major resonance contributors to the following.

a.

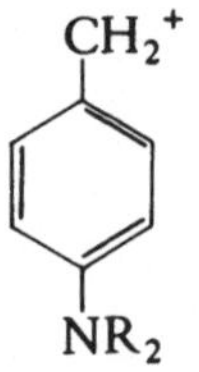

b.

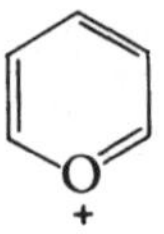

c.

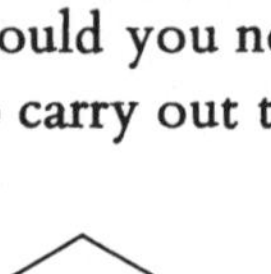

---

10. Would you need an oxidizing agent, reducing agent, or neither to carry out the following reactions?

a.

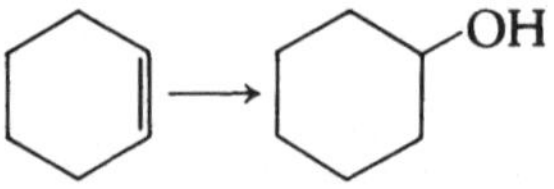

b.

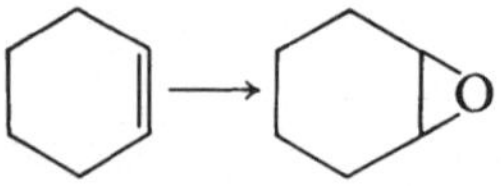

c.

d.

e.

f.

g.

h.

---

11. With each pair of compounds below indicate whether the first or second is the more acidic.

a.

b. $CH_3-\overset{\overset{\displaystyle O}{\|}}{C}-CH_3$ $CH_3-\overset{\overset{\displaystyle O}{\|}}{C}-CH_2-\overset{\overset{\displaystyle O}{\|}}{C}-CH_3$

c. $(CH_3)_3\overset{+}{N}-CH_2-COOH$ $(CH_3)_2N-CH_2-COOH$

d. $(C_6H_5)_3CH$ $(C_6H_5)_2CH_2$

e. $(CH_3)_3COH$ $(CH_3)_2CHOH$

---

12. With each pair of compounds below indicate whether the first or second is the more basic.

a. $CH_3-\overset{\overset{\displaystyle O}{\|}}{C}-NR_2$ $CH_3-\overset{\overset{\displaystyle O}{\|}}{C}-CH_3$

b. CH_3OH C_6H_5OH

c.

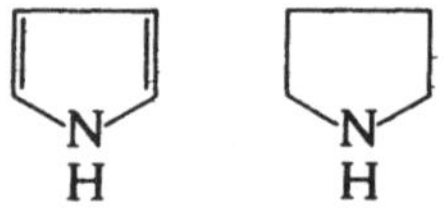

d. $(C_6H_5)_3C^-$ $(C_6H_5)_2CH^-$

---

13. Give all the reasonable resonance contributors of the following compound.

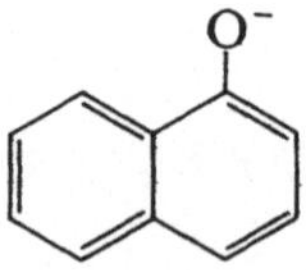

14. Is there "direct" resonance interaction between the two substituents in the following compounds?

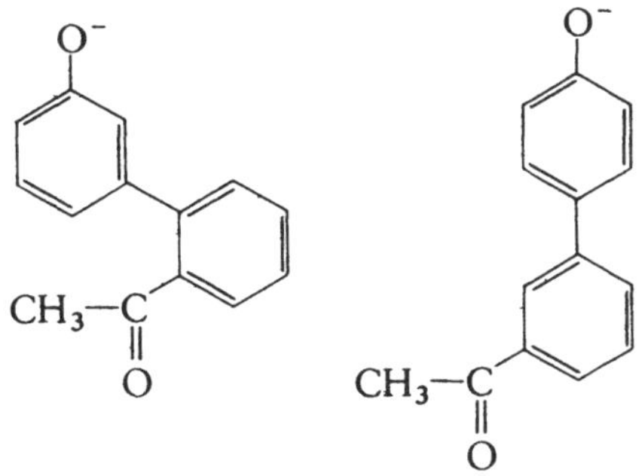

1.9. Answers

1.	a. 2	e. 1	i. 1
	b. 1	f. 0	j. 0
	c. 3	g. 1	k. 1
	d. 1	h. 2	l. 0

2. a. positive
   b. negative
   c. positive
   d. neutral
   e. oxygen, negative; sulfur, positive;
   f. positive
   g. neutral
   h. negative
   i. neutral

---

3. a. trivalent, none
    b. trivalent, none
    c. monovalent, three
    d. divalent, two
    e. trivalent, one

---

4. a. The *ortho* compound is more acidic because the bromine, which inductively withdraws electrons away from carboxyl proton, is closer to the carboxyl than in the *para* isomer. The inductive effect decreases with increasing distance.

    b. The pK_a values indicated are those for the conjugate acid of the amines. Since the conjugate acid of the *meta* compound is the stronger acid, the corresponding free amine is the weaker base. This is reasonable because the electronegative chlorine is closer to the amino group in the *meta* compound.

5.

a.

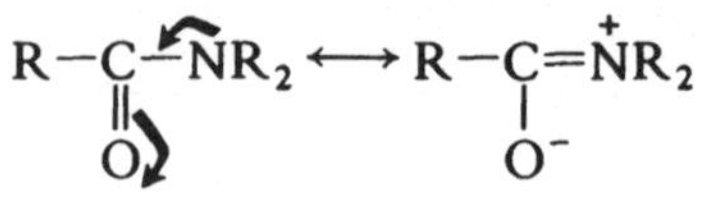

b.

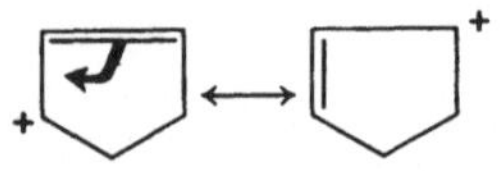

c.

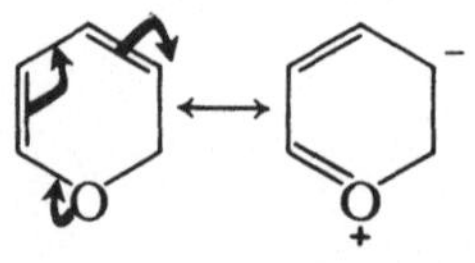

d.

e.

f.

6.

a.

b.

c.

d. $R_2\overset{+}{N}=$⟨⟩$=N-O^-$

e.

f.

Note: In structure f it is not possible to place the negative charge onto the keto group in the four-membered ring. This *would* become feasible if the OH were situated on carbons 1 or 3.

---

7. a. same compound
   b. resonance contributors
   c. same compound but different conformations
   d. resonance contributors
   e. isomers
   f. isomers
   g. isomers
   h. isomers
   i. isomers

Note: The special type of isomerism in structures f is given the name "valence bond tautomerism"; the geometries of the two species are different and thus are not resonance contributors.

8. a. Resonance contributors entail mobile electrons only (i.e., π electrons and unshared pairs). Moreover, there is no reason why one of the two carbons should possess both electrons of the σ bond while the other becomes electron deficient.

   b. The second species has a carbon with 10 valence electrons.

   c. Both fragments have different geometries than in the cyclobutene. Compare, for example, the linear H—C—C—H arrangement in acetylene with the bent arrangement of the same four atoms in cyclobutene. Thus, the structures shown constitute a chemical reaction, not resonance contributors.

   d. Electron flow moves away from the more electronegative oxygen producing a positively charged oxygen with only six outer electrons.

   e. Allenic structures (A=B=C) are linear. Thus, the right-hand structure violates Rule 6 on resonance, which states that contributors must have reasonable bond lengths and angles.

9.

a.

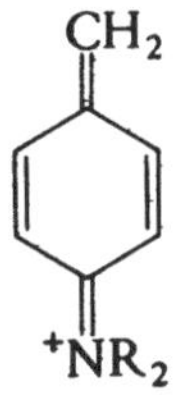

b.

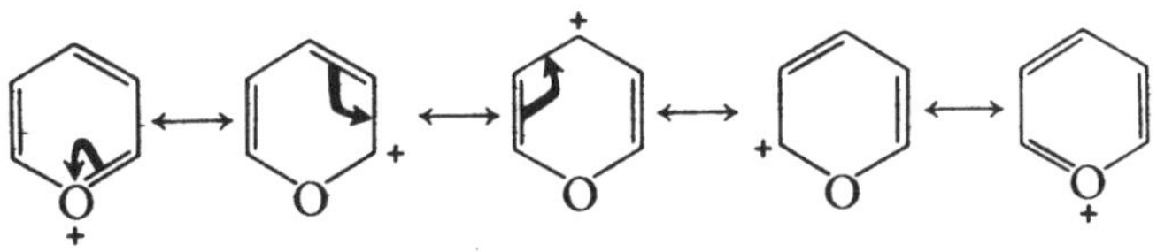

c.

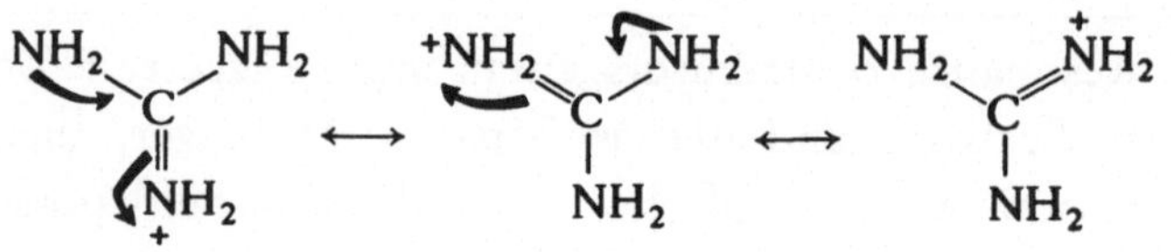

10. a. neither
   b. oxidizing agent
   c. oxidizing agent
   d. reducing agent
   e. neither
   f. neither
   g. neither
   h. reducing agent

11. a. second
   b. second
   c. first
   d. first
   e. second

12. a. first
   b. first
   c. second
   d. second

Note: The amide in problem (a) protonates (in concentrated H_2SO_4) on the oxygen to give a resonance-stabilized cation.

13.

---

14. No. There is no way of "pushing" the negative charge on the
    phenolate oxygen onto the oxygen of the ketones.

CHAPTER 2

Solved Problems in Organic Reaction Mechanisms

1.

$$CH_3O^- + H^+ \longrightarrow CH_3OH$$

A curved arrow will be used throughout this book to symbolize electron movement. For the sake of avoiding confusion it is important to have the arrows go from regions of high electron density to regions of low electron density:

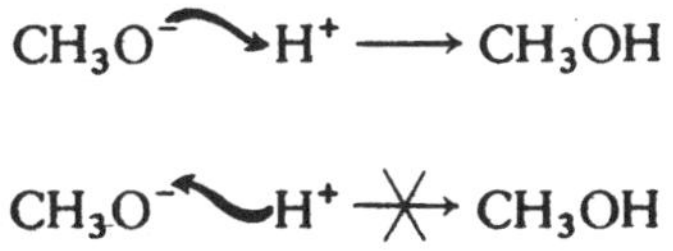

$$CH_3O^{\frown}H^+ \longrightarrow CH_3OH \quad \text{(correct)}$$

$$CH_3O^{\frown}H^+ \xrightarrow{\;\;\times\;\;} CH_3OH \quad \text{(incorrect)}$$

2.

$$HO^- + CH_3-Br \longrightarrow CH_3-OH + Br^-$$

The methyl carbon bears a partial positive charge because of the presence of the electronegative bromine. Hence, hydroxide ion (and other nucleophiles) attack the carbon and displace the bro-

---

mide in a typical S_N2 process:

$$HO^- \curvearrowright CH_3 \curvearrowright Br \longrightarrow HO-CH_3 + Br^-$$

3.
$$HO^- + (CH_3)_3C-Br \longrightarrow (CH_3)_3C-OH + Br^-$$

The three methyl groups on *t*-butyl bromide sterically impede the direct attack by hydroxide ion. On the other hand, the methyl groups favor the unimolecular cleavage of the C—Br bond by stabilizing the ensuing carbonium ion. (Recall that methyl groups are somewhat electron donating.) Thus, the mechanism involves S_N1 heterolysis, heterolysis being defined as bond cleavage in which one atom retains *both* electrons of the original covalent bond.

$$CH_3-\underset{\underset{CH_3}{|}}{\overset{\overset{CH_3}{|}}{C}}\curvearrowleft Br \longrightarrow CH_3-\underset{\underset{CH_3}{|}}{\overset{\overset{CH_3}{|}}{C^+}}\curvearrowleft OH \longrightarrow CH_3-\underset{\underset{CH_3}{|}}{\overset{\overset{CH_3}{|}}{C}}-OH$$

The substitution reaction takes place in *two* steps with a carbonium intermediate. An intermediate is a species, often of high energy, which is formed during the course of a reaction. Occasionally an intermediate is so stable it can be isolated, but in the above case the carbonium ion has a fleeting existence and is found only in low steady-state concentrations.

4.

This reaction is most likely an E2 elimination in which the basic R_3N attacks the proton, thus releasing a pair of electrons which can establish the carbon–carbon double bond by ejecting the bromide:

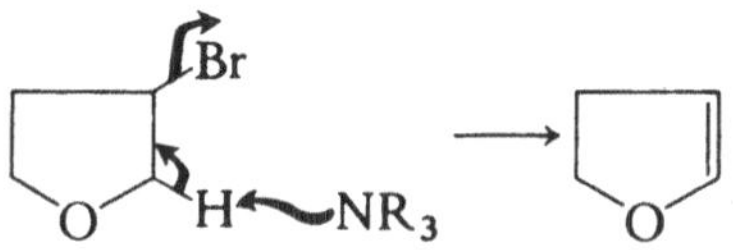

Why is the product a more likely one than its isomer shown below?

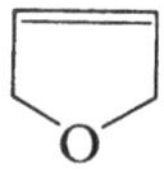

5.

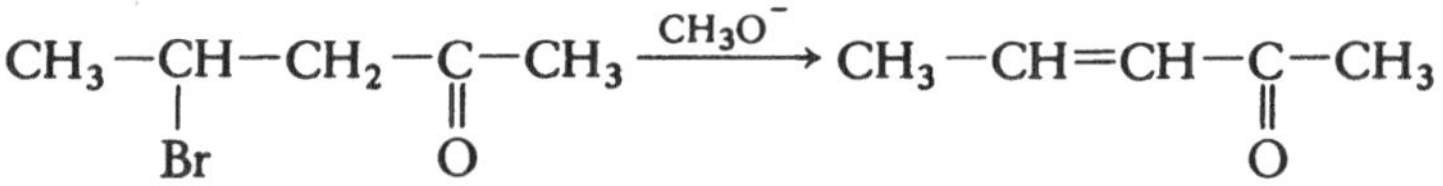

$$CH_3-\underset{\underset{Br}{|}}{CH}-CH_2-\underset{\underset{O}{\|}}{C}-CH_3 \xrightarrow{CH_3O^-} CH_3-CH=CH-\underset{\underset{O}{\|}}{C}-CH_3$$

This is a particularly favorable elimination reaction; an enolate anion intermediate may precede ejection of bromide:

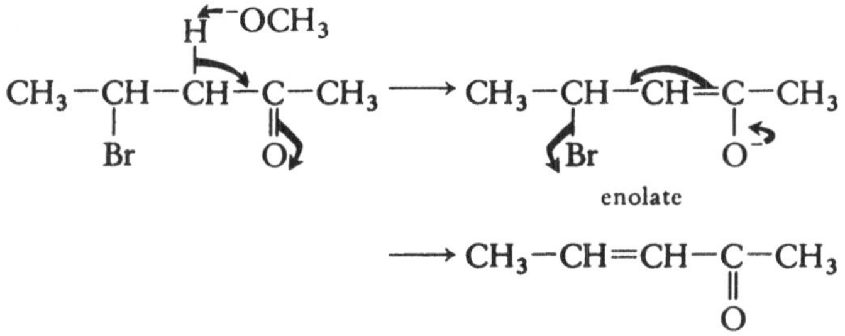

enolate

$$\longrightarrow CH_3-CH=CH-\underset{\underset{O}{\|}}{C}-CH_3$$

6.

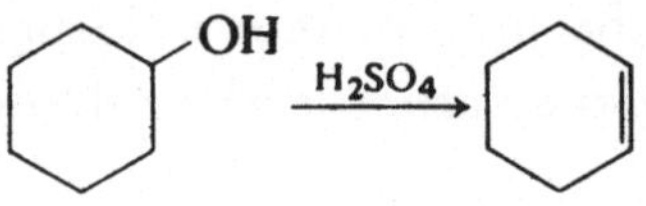

Hydroxy and alkoxy functionalities are extremely poor leaving groups in elimination and substitution reactions:

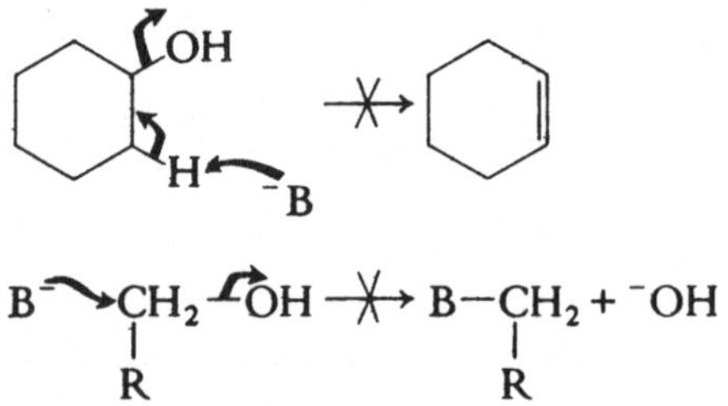

(impossible)

B^-CH$_2$—OH $\xrightarrow{\times}$ B—CH$_2$ + $^-$OH   (impossible)
     | |
     R   R

The purpose of the H$_2$SO$_4$ in the above elimination is to convert the alcohol into a species with a much better leaving group (water):

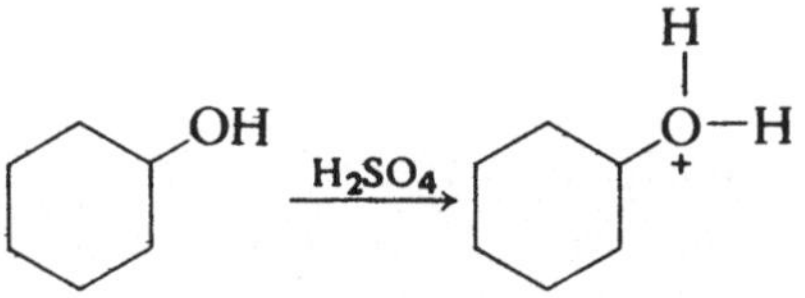

Placing a positive charge on the oxygen enhances the polarization of the electrons in the C—O bond and permits the elimination of water:

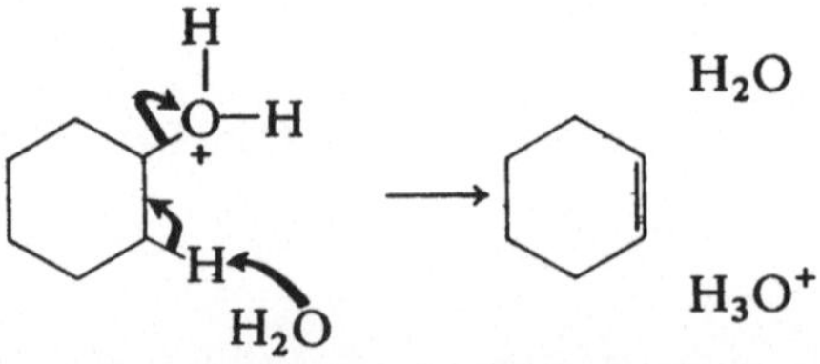

Two points are worthy of note here. First, although an acid is necessary for the reaction, a proton is neither consumed nor generated; the acid simply serves as a *catalyst*. Second, the elimination is seen to involve *water* as the base. It would have been an error to postulate hydroxide ion as the base because the reaction is run in strong acid where the hydroxide ion concentration is negligible. This brings to mind a fundamental rule: pay attention to the reaction conditions so as to avoid the use of unlikely species in the mechanism. For example, if the reaction is run under nitrogen, do not use O_2; if the reaction is run in strong aqueous base, do not use hydronium ion; if the reaction is run under weakly basic conditions, do not postulate a highly unstable carbanion.

7.

$$\underset{\underset{Br}{|}}{\overset{\overset{Br}{|}}{CH_3-CH-CH_2}} \xrightarrow{\text{Zn}} CH_3-CH=CH_2$$

This reaction entails the cleavage of two C—Br bonds (supplying a total of four electrons) and the formation of one C—C double bond (consuming two electrons). The extra two electrons are undoubtedly retained by one of the bromines as it is converted into bromide ion, but this leaves one of the bromines electron-deficient:

$$\underset{\underset{Br}{}}{\overset{\overset{Br}{}}{CH_3-CH-CH_2}} \longrightarrow \overset{Br^+}{CH_3-CH=CH_2}$$
$$\quad\quad\quad\quad\quad\quad\quad\quad\quad Br^-$$

Electron-deficient bromine (Br^+) is an unstable species, which explains in part why 1,2-dibromides do not spontaneously form alkenes. Thus far our analysis of the debromination has ignored an important component of the reaction, namely, metallic zinc. Metals are strong electron donors, and zinc can rid itself of two electrons to achieve a filled shell configuration:

$$Zn \longrightarrow Zn^{2+} + 2e^-$$

In the zinc-mediated elimination, the two electrons are supplied to one of the bromines to form zinc bromide:

$$Zn:\curvearrowright Br$$
$$CH_3-CH-CH_2 \longrightarrow CH_3CH=CH_2 + ZnBr_2$$
$$\quad\quad\quad Br$$

8.

$$CH_3-CH=CH_2 \xrightarrow{HBr} CH_3-CH-CH_3$$
$$\quad\quad\quad\quad\quad\quad\quad\quad Br$$

Which of the two components of HBr (H^+ or Br^-) would be expected to react first with a double bond? The answer is H^+ because this positively charged species has an affinity for electron-rich functionalities such as the π bond:

$$CH_3-CH=CH_2 \longrightarrow CH_3-CH-CH_3$$
$$\quad\quad\quad\quad H^+ \quad\quad\quad\quad\quad +$$

In contrast, alkenes will not react with bromide (as supplied, for example, by sodium bromide):

$$CH_3-CH=CH_2 + Br^- \longrightarrow \text{no reaction}$$

Note that proton bonding to the terminal carbon is to be preferred over bonding to the central carbon since the latter process gives a more unstable primary carbonium ion:

$$CH_3-CH{=}CH_2 \xrightarrow{\quad\times\quad} CH_3-CH_2-CH_2{}^+$$
$$\underset{H^+}{}$$

In any event, the final product forms when the secondary carbonium ion intermediate picks up a bromide:

$$CH_3-\overset{+}{C}H-CH_3 \longrightarrow CH_3-CH-CH_3$$
$$\underset{Br^-}{} \qquad\qquad\qquad \underset{Br}{|}$$

The two-step formation of 2-bromopropane from propene is a *reasonable* mechanism; that is, it obeys the rules and laws of organic chemistry. Whether or not the mechanism is *correct* is another matter; this can be ascertained only by an experiment. One could have postulated, for example, a one-step addition of HBr:

$$CH_3-CH{=}CH_2 \longrightarrow CH_3-CH-CH_3$$
$$\underset{Br-H}{} \qquad\qquad\qquad \underset{Br}{|}$$

Although this mechanism cannot in fact occur (as we shall see later on in this book when orbital considerations are discussed), at the moment we cannot *a priori* dismiss it.

9.

In the previous mechanism we alluded to the fact that nucleophiles (bromide, hydroxide, amines, cyanide, etc.) fail to react
with carbon–carbon double bonds. This is not the case with
carbon–oxygen double bonds, which are polarized and susceptible
to nucleophilic attack at the carbon:

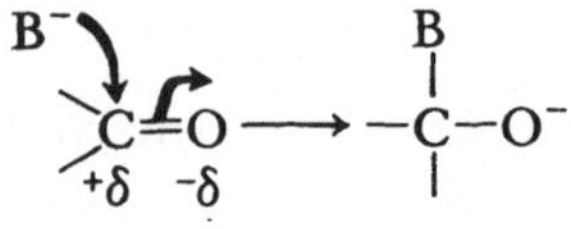

Addition to carbonyls is one of the most important reactions in
organic chemistry. When HCN is added as in the above example,
the carbonyl first becomes protonated, thereby further enhancing
reactivity toward the nucleophilic cyanide:

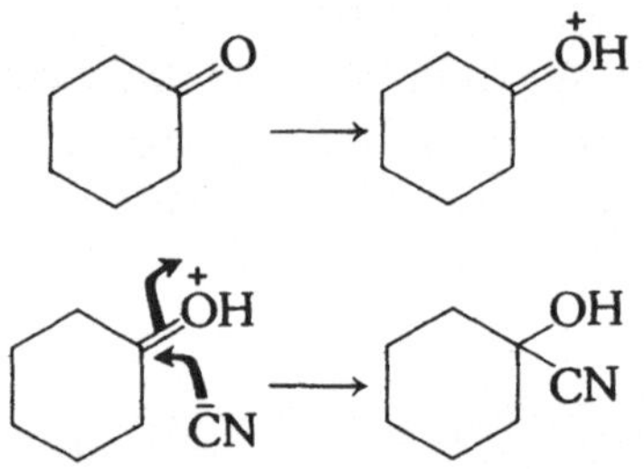

In writing reaction mechanisms, the timing of the proton bonding
to the oxygen (i.e., whether or not the proton bonds prior to or
simultaneously with cyanide addition) is merely a detail. Thus, we
can combine the two steps into one:

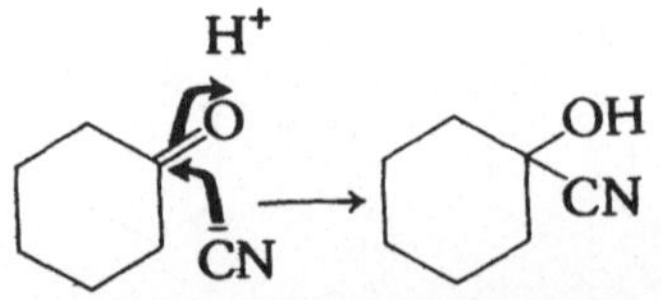

10.

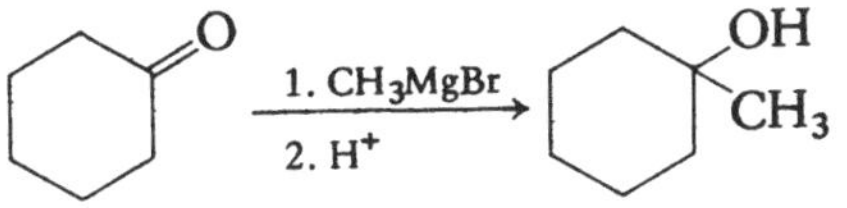

The Grignard reaction, one of only three reactions whose discovery merited the Nobel prize, is mechanistically nothing more than addition of a nucleophile to a carbonyl. The nucleophile in this case is a carbanion formed when magnesium metal reacts with an alkyl halide:

$$CH_3 - Br \quad :Mg \longrightarrow CH_3^- + \overset{+}{M}gBr$$

Actually, the carbanion is not really "free" but instead complexed with the magnesium. Thus the Grignard has both covalent and ionic character:

$$CH_3 - MgBr \longleftrightarrow \bar{C}H_3 \ \overset{+}{M}gBr$$

Whatever the exact structure of the Grignard reagent in solution, one can view it as a source of carbanions:

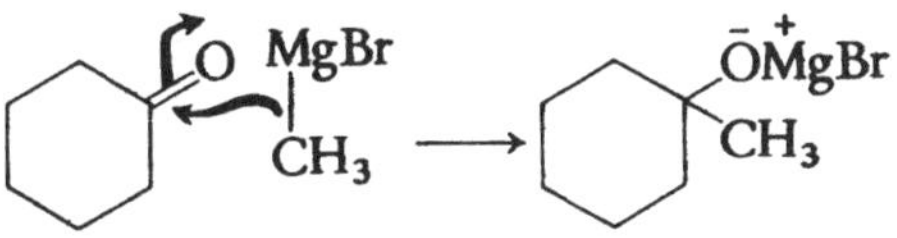

The initial product is the magnesium salt of the tertiary alcohol: when the substance is exposed to acid in the work-up (i.e., product isolation), the salt is converted into the alcohol:

$$RO^- \ \overset{+}{M}gBr \xrightarrow{\text{HX}} ROH + MgBrX$$

The above step is analogous to a neutralization reaction, NaOH + HCl gives HOH + NaCl. Note that acid is not added until the addition reaction is completed because acid would destroy the Grignard reagent:

$$H^+ \overset{\frown}{\quad} \overset{-}{C}H_3\overset{+}{M}gBr \longrightarrow CH_4$$

11.

One sees that the components are available for an addition reaction: carbonyl group, nucleophile (the amino group of hydroxylamine), and an acid catalyst:

Loss of a proton from the cationic nitrogen gives the addition product:

Since proton transfers to and from heteroatoms such as nitrogen and oxygen are frequently very fast low-energy processes, we will in the future not write the above type of reaction in two steps but rather combine them:

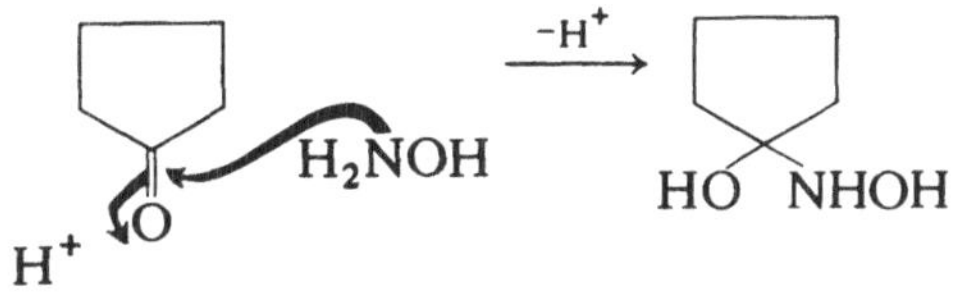

For our purposes the two descriptions are equivalent, but brevity favors the second. The reaction, however, does not terminate here. Water is eliminated from the addition product to give the oxime:

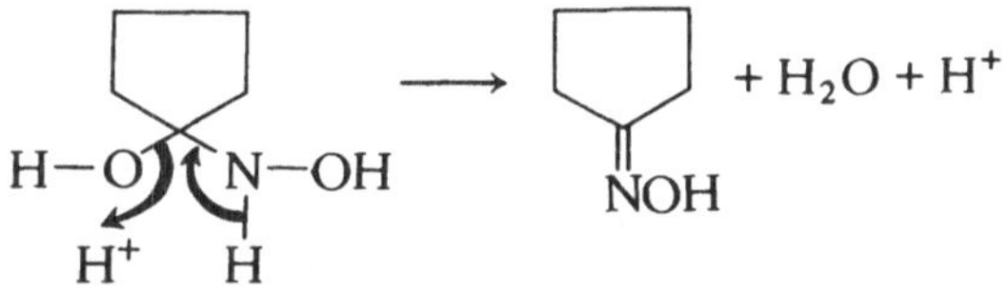

The above step is mechanistically no different from the dehydration of an alcohol:

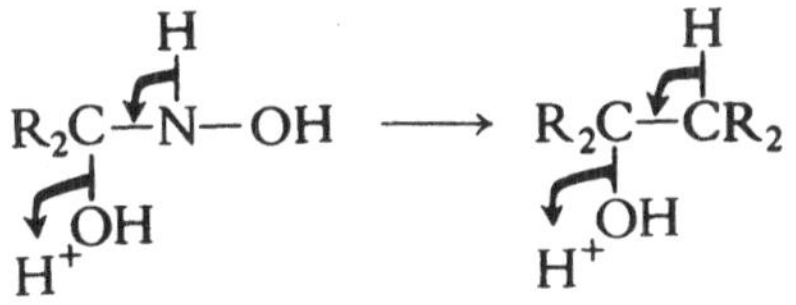

12.

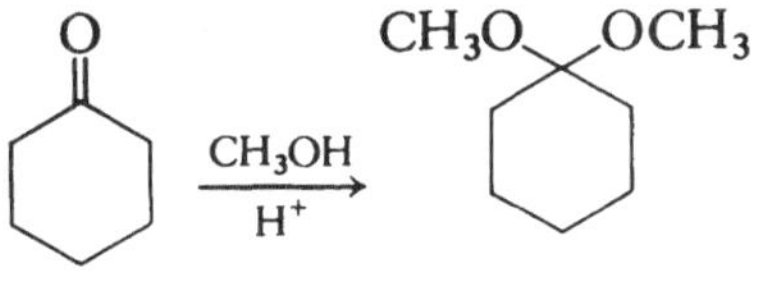

Conversion of a ketone into a ketal involves another addition-elimination reaction:

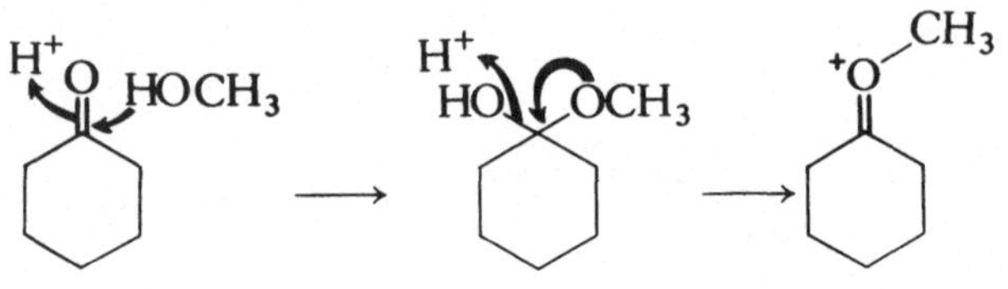

Addition of a second molecule of methanol to the oxonium ion intermediate produces the ketal:

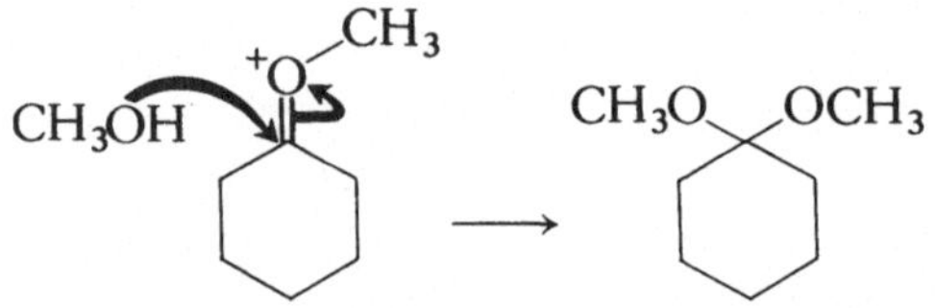

One may ask why the ketal could not be formed directly via an S_N2 displacement reaction on the addition intermediate:

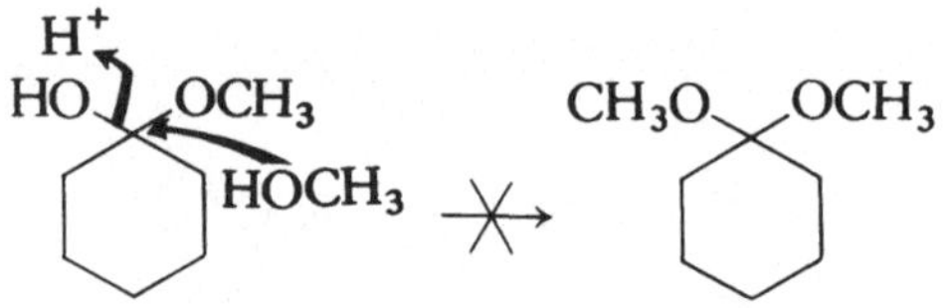

Three factors make such a process unlikely: (a) Methanol is a poor nucleophile in S_N2 reactions (in contrast to methoxide ion). (b) Severe steric effects at the reactive carbon would impede the substitution (halides follow the order primary > secondary > tertiary in S_N2 reactivity). (c) There is little reason why water should be ejected using electrons from an external methanol molecule when electrons are available from an oxygen already on the addition intermediate.

All the reactions in the ketal formation are reversible. Thus, it is possible to hydrolyze a ketal back to the ketone and alcohol in aqueous acid. Although in methanol the reaction is shifted toward the ketal, in water the ketone is favored. As an exercise, the student should write the mechanism for the hydrolysis of a ketal.

13.

$$\text{(pyrrolidine N-H)} + CH_2{=}CH{-}\underset{\underset{O}{\|}}{C}{-}CH_3 \longrightarrow \text{(pyrrolidine N)}{-}CH_2{-}CH_2{-}\underset{\underset{O}{\|}}{C}{-}CH_3$$

The α,β-unsaturated ketone has two potential sites for nucleophilic attack—the carbonyl carbon and the β carbon:

$$\overset{\beta}{C}H_2{=}\overset{\alpha}{C}H{-}\underset{\underset{O}{\|}}{C}{-}CH_3 \longleftrightarrow CH_2{=}CH{-}\overset{+}{C}{-}CH_3 \longleftrightarrow \overset{+}{C}H_2{-}CH{=}\underset{\underset{O^-}{|}}{C}{-}CH_3$$

No doubt the amine adds to both centers, but addition to the carbonyl gives an intermediate which can revert back to the ketone. In other words, addition is reversible:

$$CH_2{=}CH{-}\underset{\underset{O}{\|}}{C}{-}CH_3 \rightleftarrows CH_2{=}CH{-}\underset{\underset{O^-}{|}}{C}{-}CH_3$$

In contrast, amine attack at the β carbon gives an enolate ion which picks up a proton to form the ketone. This latter step drives the amine addition to the right:

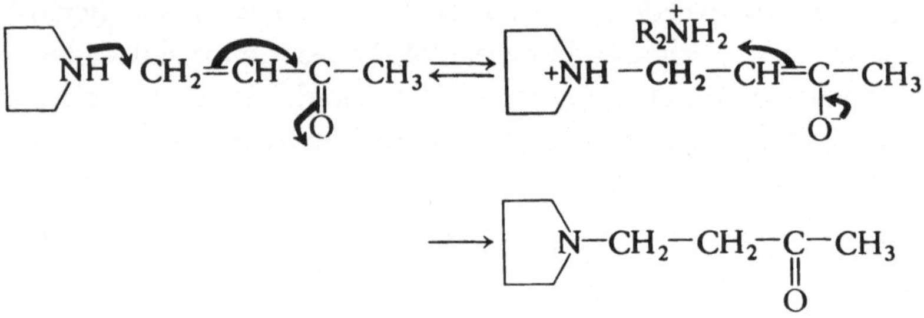

The proton source is probably a protonated amine (H_3O^+ would not be found in the highly basic reaction medium). In summary, addition at the carbonyl is a "blind alley" which does not prevent so-called *Michael addition* at the β carbon; ultimately all the starting material is converted into Michael product.

14.

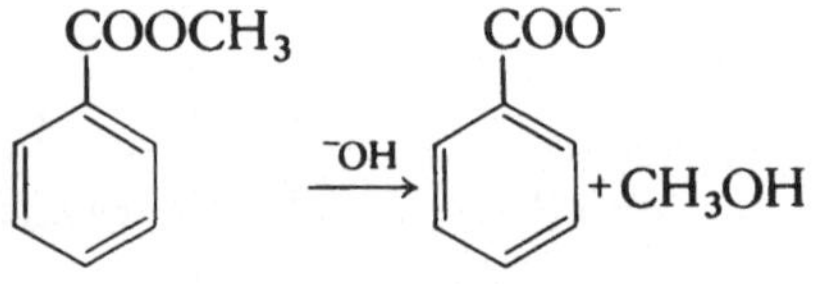

The mechanism for basic ester hydrolysis is known to be that shown below:

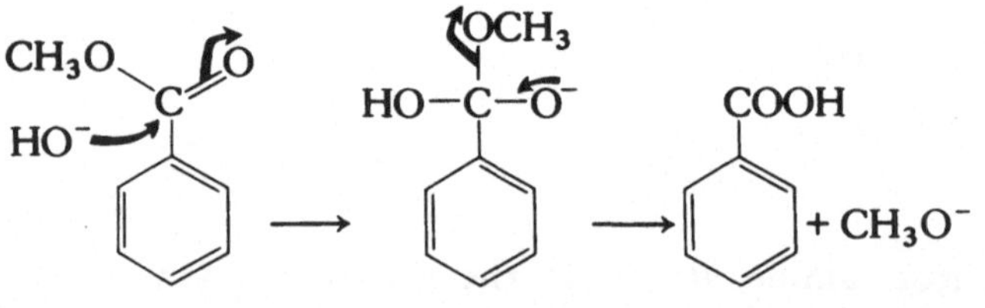

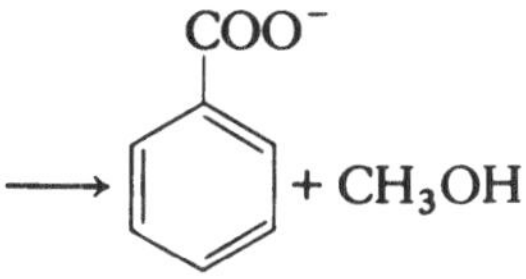

$$\longrightarrow \quad \boxed{\bigcirc} + CH_3OH$$

COO⁻ attached to benzene ring + CH₃OH

Although this is not a text in physical organic chemistry, it might be well at this point to show that the mechanism is not only *reasonable*, it is also in essence *correct*. As far as being reasonable, the mechanism is nothing more than an addition–elimination process which we have encountered previously. An alternative but incorrect mechanism is shown below:

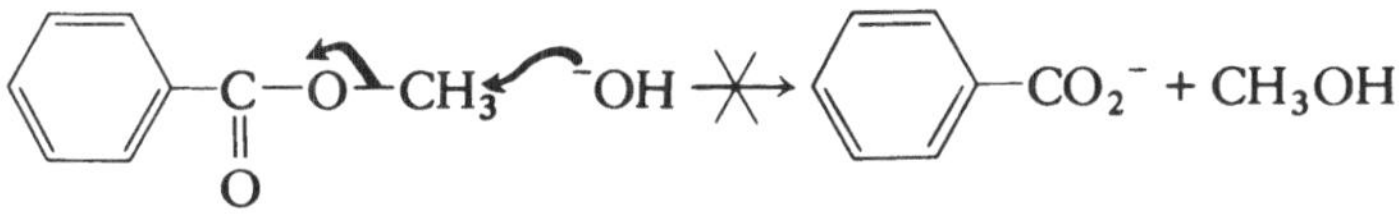

The direct substitution route seems unappealing because benzoate is a poor leaving group in S_N2 reactions. Experiments with ^{18}O isotope labeling leave no doubt that hydrolysis arises from attack at the carbonyl carbon and not the methyl group. An ester labeled with ^{18}O at the "ether" oxygen was hydrolyzed in base, and the resulting alcohol was isolated and examined for its ^{18}O content. The amount of ^{18}O in the alcohol was exactly the same as the amount in the original ester:

$$R-\underset{\underset{O}{\|}}{C}-^{18}O-R' \xrightarrow{\ ^-OH\ } R-\underset{\underset{O}{\|}}{C}-O^- + H-^{18}O-R'$$

$$(\text{no }^{18}O)$$

If hydroxide ion attack had occurred at the alkyl group, isotopic oxygen would have been found in the acid portion of the products.

Hydrolysis via carbonyl attack might conceivably occur either by direct displacement or by an addition–elimination mechanism:

(A) Direct displacement

$$HO^- \searrow$$
$$R-\overset{\|}{\underset{O}{C}}\!\!\!\!\!\rightarrow OR \longrightarrow \text{products}$$

(B) Addition–elimination

$$HO^- \searrow \qquad\qquad OH$$
$$R-\overset{\|}{\underset{O}{C}}-OR \longrightarrow R-\overset{|}{\underset{O^-}{C}}\!\!\!\nearrow OR \longrightarrow \text{products}$$

The mechanisms are quite different. The first is analogous to a simple S_N2 displacement and entails no intermediate. In the second mechanism (the correct one), the ester initially forms an unstable *tetrahedral intermediate.* This intermediate would not be expected to build up to any appreciable concentration. Since isolation of the intermediate is impossible, more subtle means must be used to prove its existence. One such proof is given below.

Methyl benzoate, labeled with ^{18}O at the carbonyl, was added to aqueous base. After about 50% of the reaction was over, the base was neutralized and unreacted ester was isolated. The amount of ^{18}O per mole in the isolated ester was *less* than the amount per mole in the original ester. Isotopic oxygen gets "washed out" of the ester during the course of the reaction. Clearly, the displacement mechanism does not explain this fact for it requires that the ester remain intact until a hydroxide ion manages to displace the methoxide ion. The mechanism involving the tetrahedral intermediate, however, is consistent with loss of isotope in the ester, as shown below:

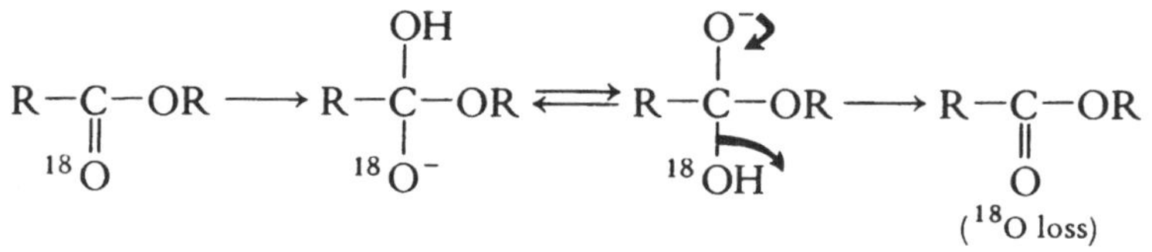

$$
\begin{array}{ccc}
\underset{\underset{^{18}O}{\overset{\|}{}}{\overset{OH}{\underset{}{}}}}{R-C-OR} \longrightarrow R-\underset{^{18}O^-}{\overset{OH}{C}}-OR \rightleftharpoons R-\underset{^{18}OH}{\overset{O\overset{\cdot\cdot}{}}{C}}-OR \longrightarrow R-\underset{O}{\overset{\|}{C}}-OR
\end{array}
$$

$$(^{18}O \text{ loss})$$

If the two hydroxy groups in the tetrahedral intermediate become equivalent through rapid proton transfer, then the isotope will be lost when the intermediate reverts back to the ester (which it will do in addition to ejecting alkoxide to form carboxylate).

It is beyond the scope of the book to give experimental support for all the mechanisms discussed. Indeed, many of the reactions that follow are complex and have eluded detailed mechanistic studies. This is not to downplay our proposed mechanisms; they illustrate plausible modes of bond breakage and bond formation as reactants are converted into products, and as such the mechanisms are very useful.

15.

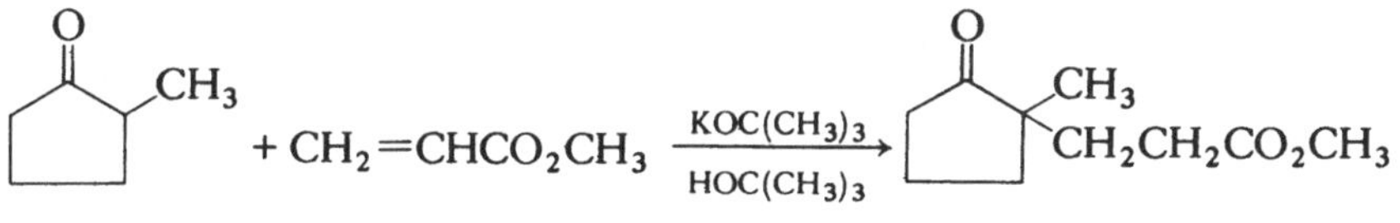

$$+ CH_2{=}CHCO_2CH_3 \xrightarrow[\text{HOC(CH}_3)_3]{\text{KOC(CH}_3)_3}$$

The strong base (potassium t-butoxide) promotes formation of the enolate:

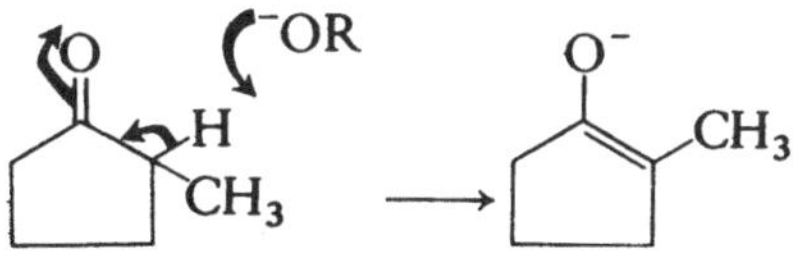

It is not necessary that the base be sufficiently strong to convert *all* the ketone to enolate; whatever enolate is present will go on to react further (the ketone serving as a reservoir to replenish the enolate). Of course, the stronger the base, the greater the concentration of enolate, and the faster the reaction. In the second step, the enolate undergoes Michael addition to the α,β-unsaturated ester:

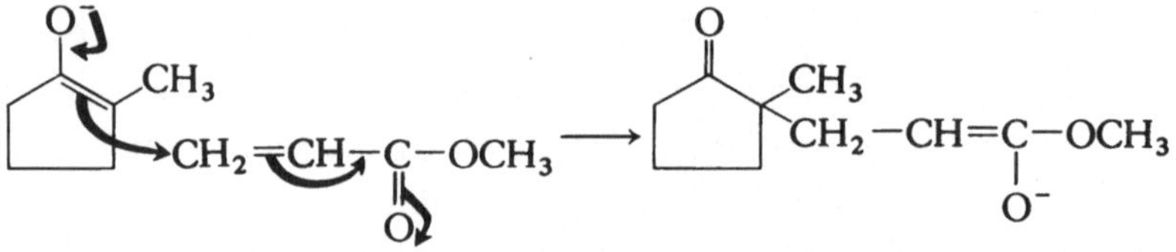

The enolate form of the ester then picks up a proton from the solvent to generate the more stable carbonyl tautomer:

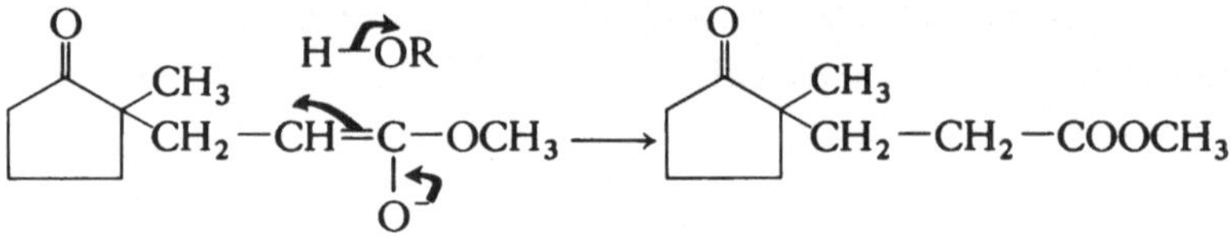

Note that the major product forms from the more substituted enolate and that C-alkylation predominates over O-alkylation (shown below):

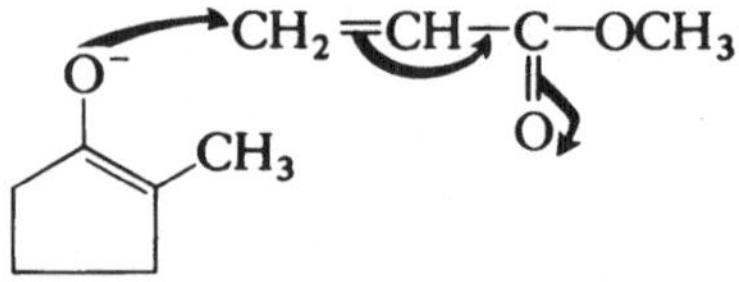

Although experience dictates that O-alkylation is a less likely

pathway, one would not be able to predict this fact from "electron pushing" alone.

16.

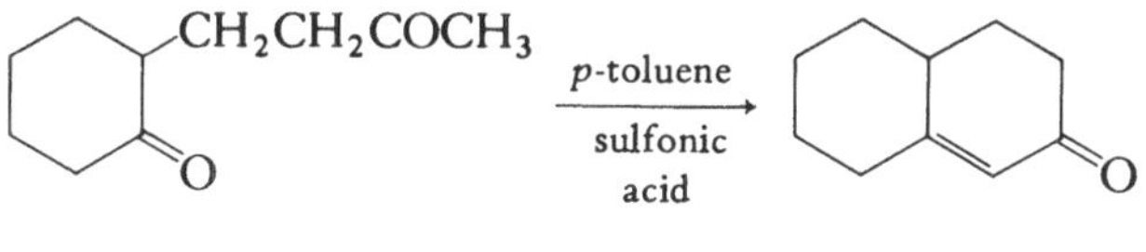

Acid-catalyzed enol formation converts the terminal methyl
group into a nucleophilic entity which adds intramolecularly to
the carbonyl. Dehydration then gives the final "aldol" product:

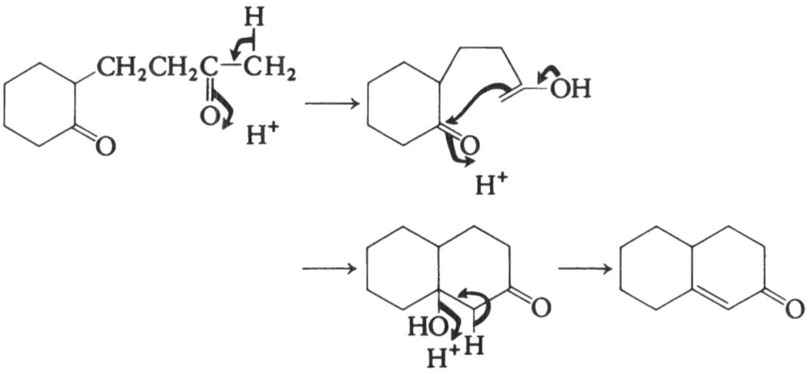

Note that an enol could, in principle, attack a carbonyl on the
same molecule or a carbonyl on *another* molecule. In general,
however, the intramolecular process predominates when a 5-
or 6-membered ring is closed. Rings with 5 or 6 atoms are particularly easy to form because the "ends" can find each other
readily and because little strain accompanies the closure. In this
connection, it is relevant that no product is observed from the

following enol:

$$\text{(cyclohexanone ring)}-CH_2CH=\underset{\underset{OH}{|}}{C}-CH_3$$

Intramolecular attack by this enol on the carbonyl would require the unlikely formation of a highly strained 4-membered ring.

In summary, the seemingly complex transformation is nothing more than an enolization, carbonyl addition, and dehydration.

17.

$$\text{(phenyl)}-CH=CH-COOH \xrightarrow[\text{heat}]{H_2SO_4} \text{(phenyl)}-CH=CH_2 + CO_2$$

There are those who feel that it is bad pedagogy to present false material. Whatever the merits of this opinion in other areas, it has no validity with reaction mechanisms; unreasonable mechanisms can be very instructive. Take, for example, the following mechanism for decarboxylation of cinnamic acid in strong acid:

$$\text{(phenyl)}-CH=CH-\underset{\underset{O}{\|}}{C}-OH \not\rightarrow \text{(phenyl)}-CH=CH^- + CO_2 + H^+$$

$$\downarrow H^+$$

$$\text{(phenyl)}-CH=CH_2$$

We can eliminate this mechanism from serious consideration for several reasons: (a) The crucial first step generates a highly unstable carbanion; the negative charge is localized on a single carbon atom since no resonance stabilization is possible. (b) The first, and certainly rate-determining, step of the mechanism makes no use of the reagent, H_2SO_4. If such a reaction were possible, cinnamic acid would be unstable and would spontaneously evolve CO_2. (c) One can think of no good analogy in organic chemistry in support of the proposed mechanism.

One might escape objection (b) by postulating the following mechanism:

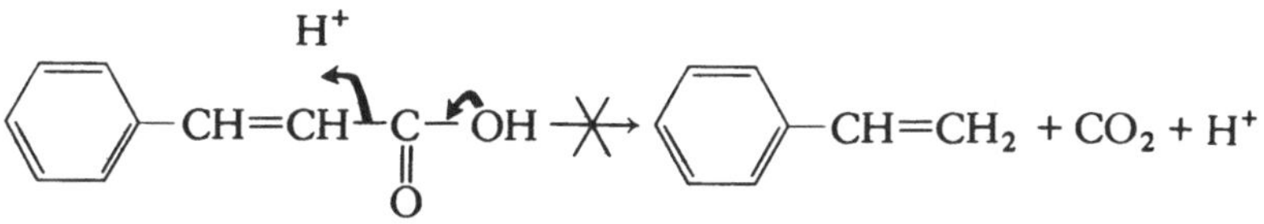

The mechanism, although incorrect, at least makes use of H_2SO_4 in the step where the carbon–carbon bond is broken. Yet the mechanism still fails in that electrons in the transition state are pushed toward an unactivated carbon (the use of a proton somewhat disguising this fact). Neither acetic acid nor ethanol undergo similar reactions in strong acid (see below), and there is no reason to suspect that cinnamic acid would behave differently.

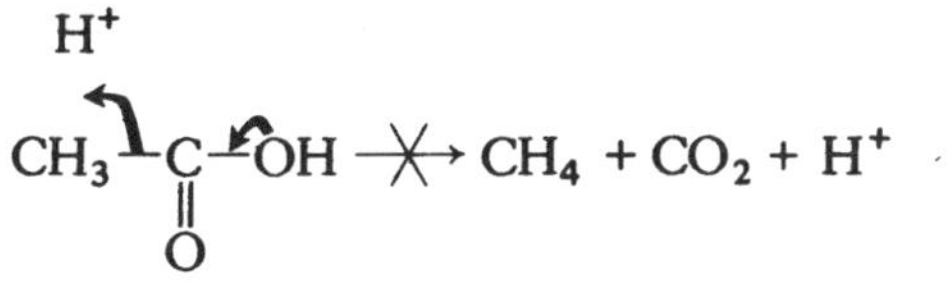

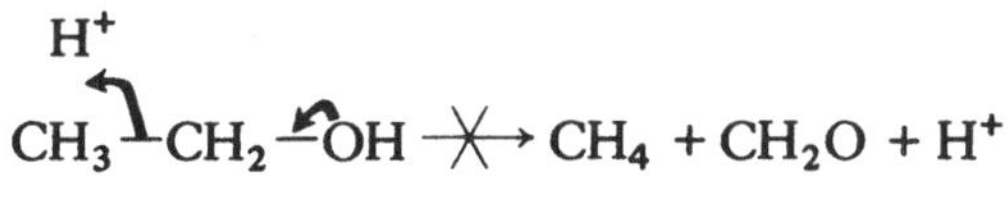

A much more reasonable sequence of events is shown below:

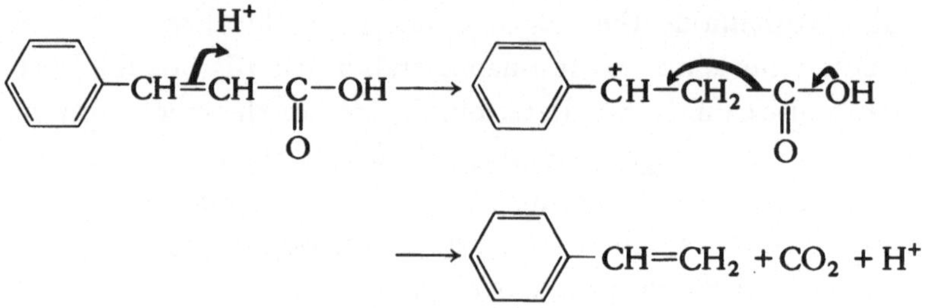

Protonation as indicated creates a resonance-stabilized carbonium ion. Clearly, this carbonium ion is much more resonance-stabilized than a carbonium ion α to a carbonyl:

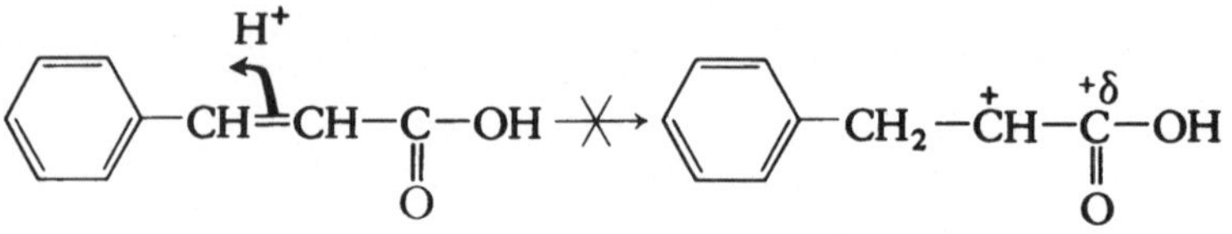

In any event, the resonance-stabilized carbonium ion serves as an electron "sink" to accept the pair of electrons released when the carbon–carbon bond is broken in the decarboxylation. As a general rule, decarboxylations are relatively facile when a suitable atom is positioned correctly to accept an electron pair. Two other examples are shown below:

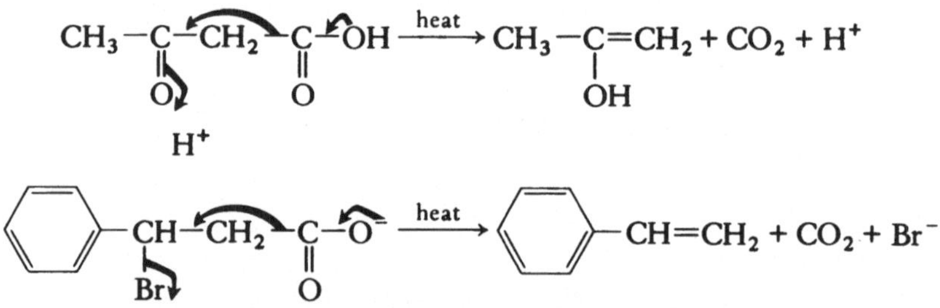

Actually, decarboxylation of acetoacetic acid probably proceeds by an *intramolecular* delivery of a proton, but the mechanism is basically the same as above:

$$\longrightarrow CH_3-\underset{\underset{OH}{|}}{C}=CH_2 + CO_2$$

Cyclic processes are often favored when the transition state contains five or six reacting atoms.

18.

$$\longrightarrow CH_3CHO + CO_2 +$$

   This is the well-known ninhydrin test for an α-amino acid. The major product has a blue color owing to a highly delocalized negative charge:

$$\longleftrightarrow$$

$$\longleftrightarrow etc.$$

A step-by-step analysis, using a series of simple reactions, makes this transformation considerably less formidable than one might at first think. A likely first step entails addition of the nucleophilic amino group on the amino acid to one of the carbonyls of the ninhydrin reagent. The structure of the product suggests that the nitrogen attacks the central carbonyl, and this is reasonable because the other two carbonyls are deactivated by resonance interactions with the aromatic ring. Elimination of water from the addition product produces an imine:

Decarboxylation now becomes electronically feasible:

In the next step, the imine hydrolyzes to the amine and aldehyde by a mechanism just the reverse of imine formation:

Finally, the amine reacts with another mole of ninhydrin in the usual manner to give an imine which, when ionized, is bright blue in color.

19.

Aromatic rings are electron-rich and, like simple alkenes, are attacked by electrophiles (H^+, Br^+, NO_2^+, etc.) *not* nucleophiles (HO^-, Br^-, R_2NH, etc.). In the above Friedel–Crafts acylation, an electrophile is generated when the acid chloride interacts with the Lewis acid, $AlCl_3$:

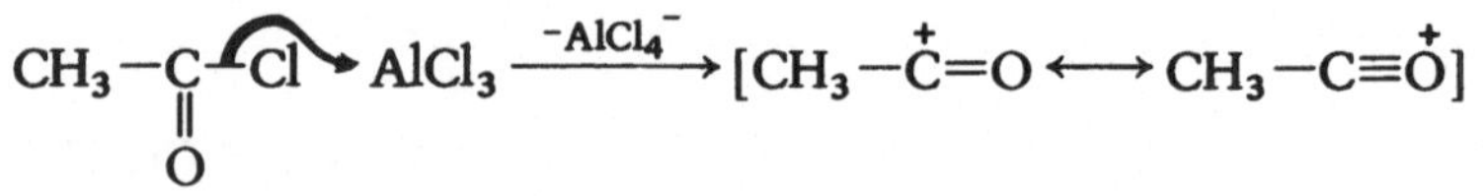

$$CH_3-\underset{\underset{O}{\|}}{C}-Cl + AlCl_3 \xrightarrow{\ -AlCl_4^-\ } [CH_3-\overset{+}{C}=O \longleftrightarrow CH_3-C\equiv\overset{+}{O}]$$

Since the Al in $AlCl_3$ has only six outer electrons, the metal can accept a pair of electrons and thus induce the chlorine to depart. The resulting *acylium ion* is a relatively stable carbonium ion owing to an important oxonium ion contributor in which all the atoms possess a full octet of outer electrons. When the acylium ion attacks the π cloud of the aromatic ring, an intermediate is formed which rapidly loses a proton to give the final product:

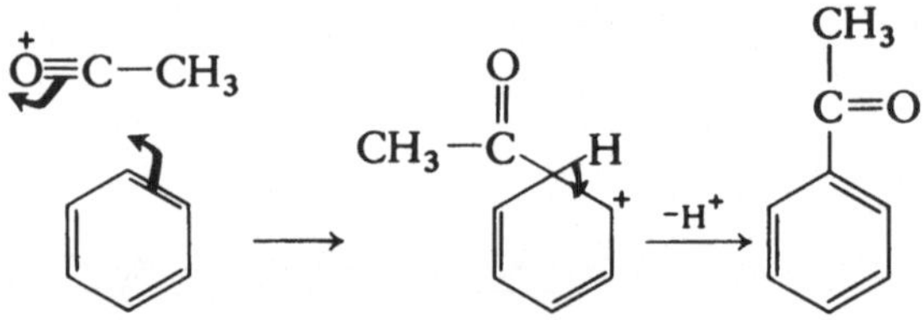

Experiments have proved the existence of the addition intermediate; a one-step mechanism apparently has a high activation energy and does not occur:

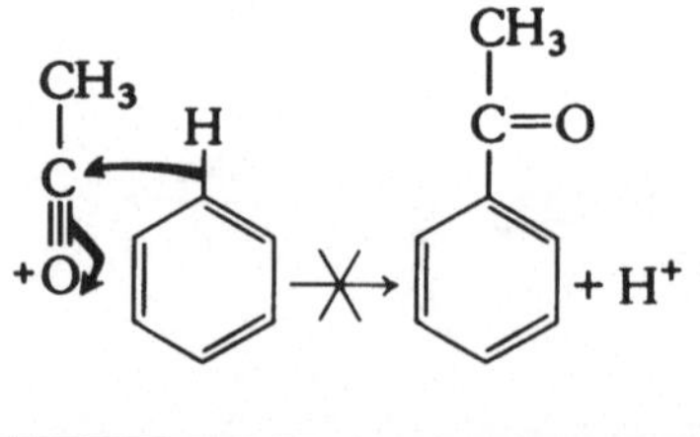

20.

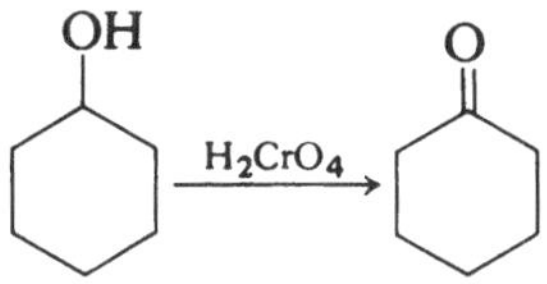

Chromate oxidation of alcohols to ketones consists of two steps:

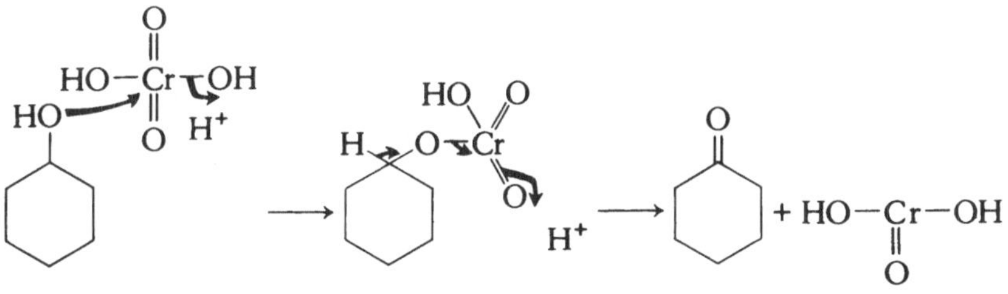

Both steps have familiar analogies. The first step resembles an S_N2 substitution, and the second step is basically an elimination reaction of the general type:

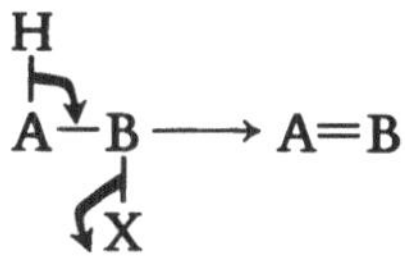

$$\overset{H}{\underset{\overset{|}{X}}{A - B}} \longrightarrow A = B$$

In the dehydration of an alcohol to an alkene, A = B = carbon and X = oxygen; in the second step of the chromate oxidation, A = carbon, B = oxygen, and X = chromium. Note that heterolytic cleavage of a Cr—O bond in the second step places *both* electrons on the metal and, as a consequence, the chromium is reduced.

21.

$$R_2C{=}O + R_2NH + HCO_2H \xrightarrow{\text{heat}} R_2CH{-}NR_2$$

In this "reductive amination" (Leuckart reaction) a ketone is converted into an amine. Comparison of the oxidation numbers of the ketone and amine reveals that the keto functionality has been reduced:

$$R_2C=O \qquad R_2CH-NR_2$$

Oxidation no.:   2   1

One immediately suspects that the third component of the reaction, formic acid, serves as the reducing agent. With this in mind, we postulate a reasonable, and by now totally familiar, first step:

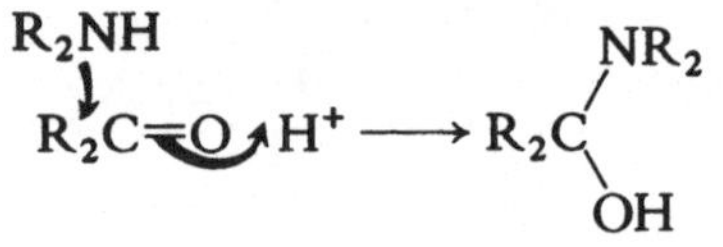

Loss of water generates the immonium ion intermediate:

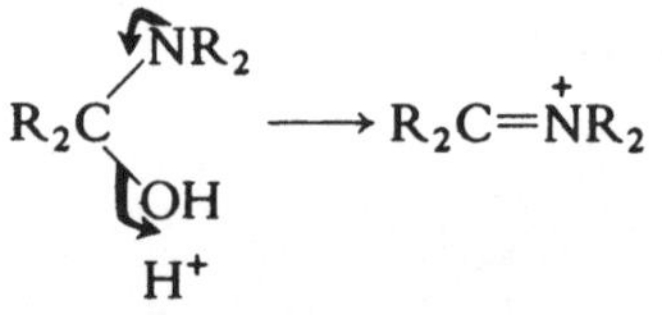

Although progress has been made in securing the final product (i.e., the crucial carbon–nitrogen bond has been formed), the intermediate retains the oxidation state of the original ketone. It is now necessary to reduce the immonium ion with formate ion:

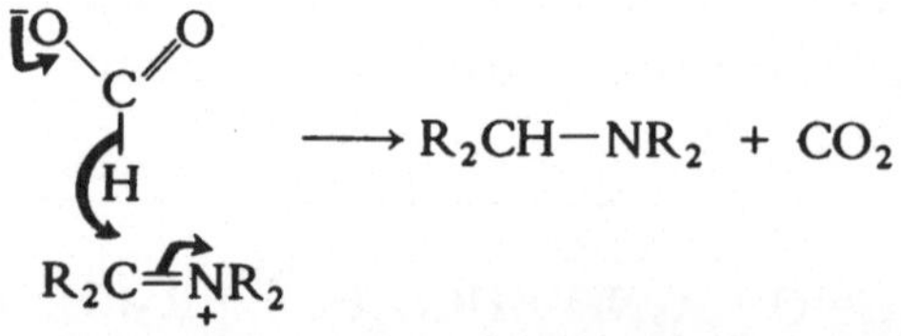

Formate ion is apparently a sufficiently good reducing agent (hydride donor) that at elevated temperatures it will reduce a ketone derivative. One may ask why the formate does not reduce the ketone itself to give a secondary alcohol. The answer must stem from the fact that the immonium ion possesses a *positive* atom as the ultimate electron acceptor, making it particularly susceptible to hydride reduction.

Most hydride reductions are carried out with inorganic hydride donors ($LiAlH_4$, $NaBH_4$, etc.); formate ion is one of only a few organic hydride donors and it is much less potent than the inorganic reducing agents. From the authors' experience, one of the most common errors of students in writing mechanisms entails unreasonable expulsion of hydride ion. Three such examples are given below:

$$R\overset{\curvearrowright}{CH}-NH_2 \overset{\times}{\longrightarrow} RCH=NH$$

None of these reactions would occur in the absence of a suitable oxidizing agent. One can determine the oxidation numbers of the reactants and products to prove that in fact the reactions constitute oxidative transformations. Unreasonable hydride expulsions are "disguised" (as in the three examples) by incomplete arrow-drawing. Using three arrows instead of two in the third "reaction" would make the absurdity more evident:

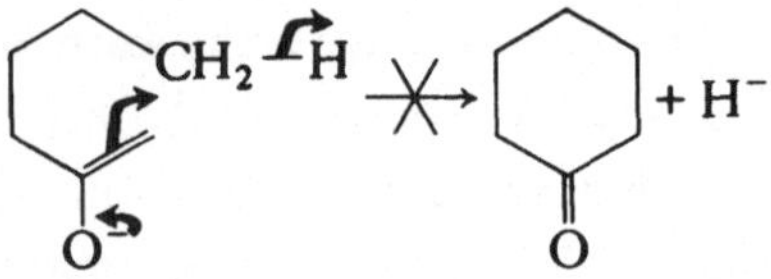

***

22.

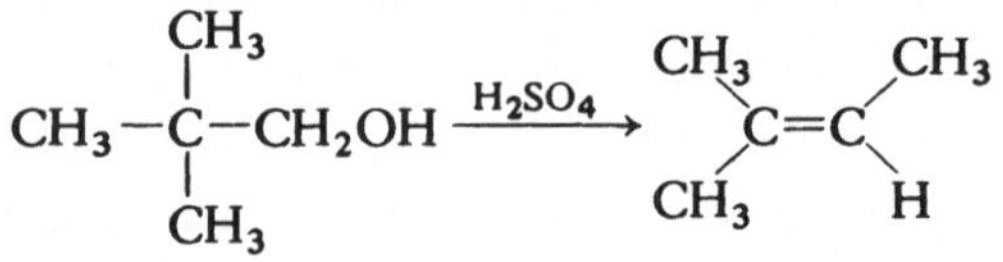

This reaction is a member of a large and important class of reactions, the rearrangements. Since rearrangements can modify the carbon skeleton of a molecule, it is not surprising that some of the most challenging mechanistic problems include a rearrangement step. A rearrangement can be best viewed as a migration of an atom or groups of atoms to an electron-poor center:

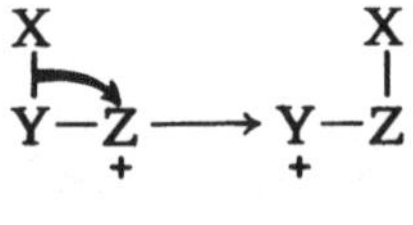

The electron-deficient atom (Z) can be carbon, oxygen, or nitrogen. Usually the driving force for the migration originates from atom Y being more able to sustain electron deficiency than Z. The Wagner–Meerwein rearrangement in the reaction of neopentyl alcohol is a classical example. The compound first protonates at the hydroxy group:

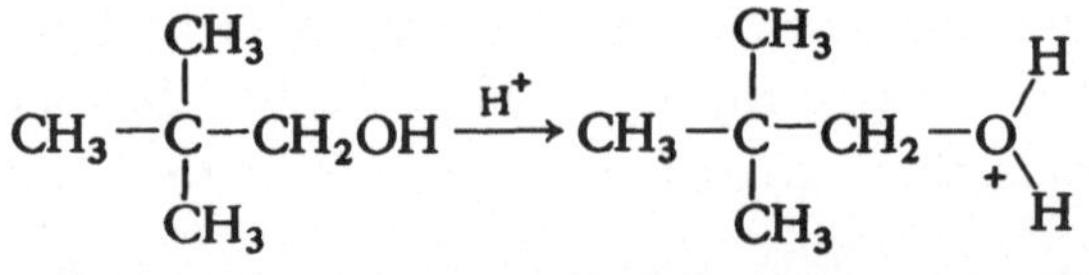

Recall that the hydroxy group is also protonated in the acid-catalyzed dehydration of alcohols. In neopentyl alcohol, however, the compound lacks the proton necessary to eliminate water. Instead, the methyl group with a pair of electrons migrates to the methylene group as the water molecule departs:

$$CH_3-\underset{\underset{CH_3}{|}}{\overset{\overset{CH_3}{|}}{C}}-CH_2-\overset{+}{O}\big\langle\begin{array}{c}H\\H\end{array}\longrightarrow CH_3-\underset{\underset{CH_3}{|}}{\overset{+}{C}}-CH_2-CH_3 + H_2O$$

It is unlikely that water leaves prior to migration because this would produce an unstable primary carbonium ion. Rearrangement, on the other hand, gives directly a relatively stable tertiary carbonium ion, accounting in part for the ease of the reaction. Proton loss gives the alkene product:

$$CH_3-\underset{\underset{CH_3}{|}}{\overset{+}{C}}-\overset{H}{CH}-CH_3 \longrightarrow CH_3-\underset{\underset{CH_3}{|}}{C}=CH-CH_3$$

In the Baeyer–Villiger oxidation, a ketone is converted into an ester by means of a rearrangement to an electron-deficient oxygen:

Study the mechanism carefully:

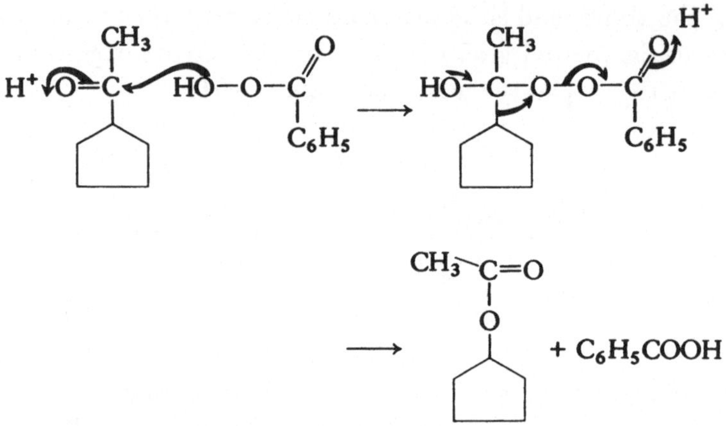

Note that the cyclopentyl group migrates in preference to the methyl group. The driving force for the rearrangement is, no doubt, the cleavage of the high-energy oxygen–oxygen bond in which two electronegative atoms are attached to each other.

23.

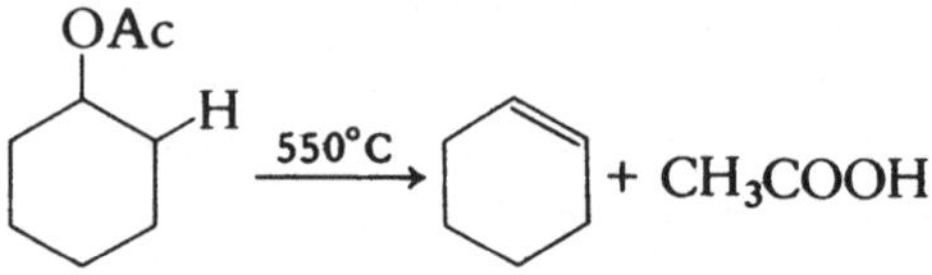

This acetate pyrolysis is carried out by heating the reactant to 550°C in a column of glass beads. Since it has been shown that the acetate group and reacting proton must be in a *cis* relationship, a cyclic *concerted* process seems likely. By "concerted" one means that bond formation and breakage occur in unison:

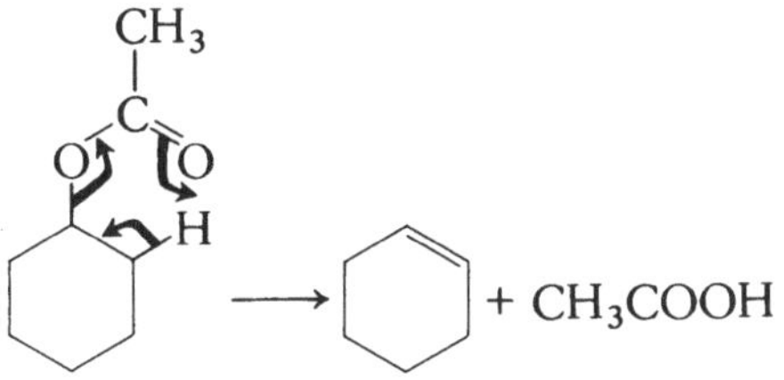

The cyclic mechanism may be favored because it precludes charge formation, as in the following alternative possibility:

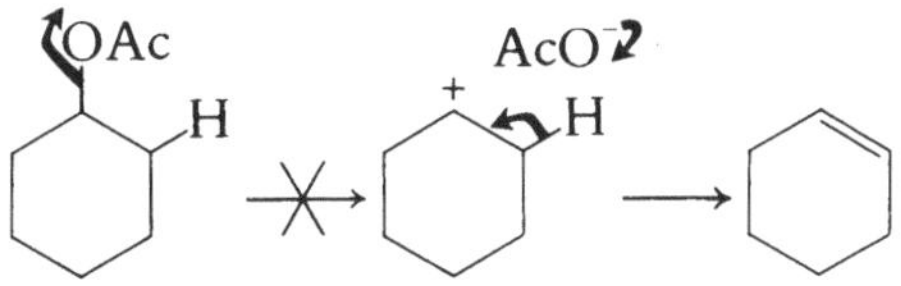

Lack of charge creation is especially important in systems where there is no solvent to solvate and stabilize ionic species.

Other examples of reactions that can be written, at least formally, as cyclic concerted processes are given below:

(a) Cope elimination

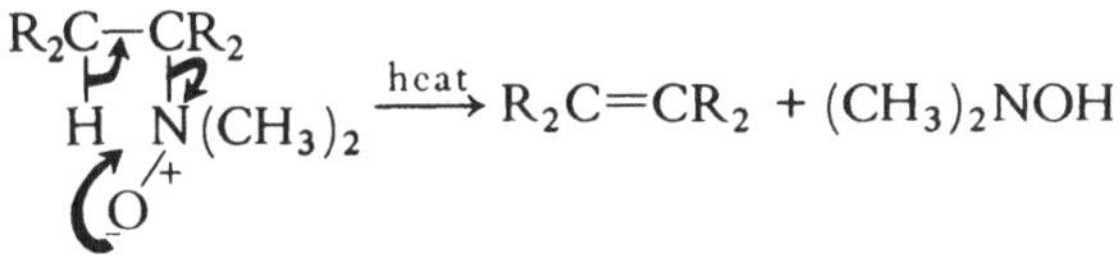

(b) Diels–Alder reaction

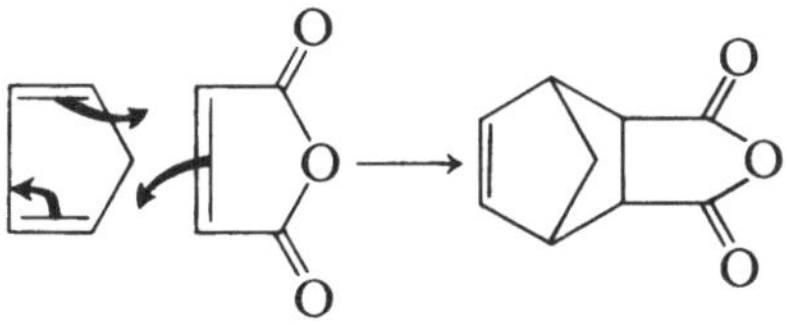

(c) *Valence tautomerism*

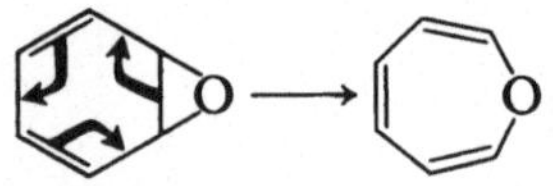

(d) *Cope rearrangement*

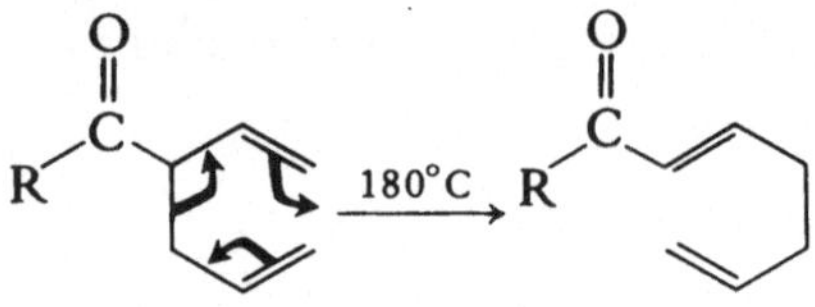

(e) *Electrocyclic rearrangement*

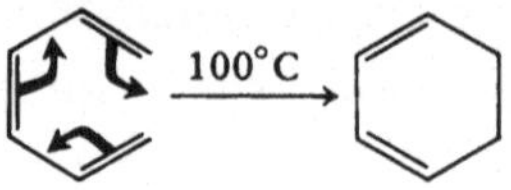

(f) *1,3-Dipolar cycloaddition*

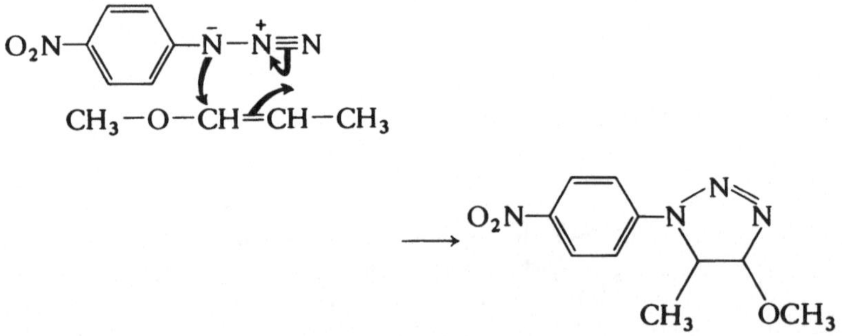

24.

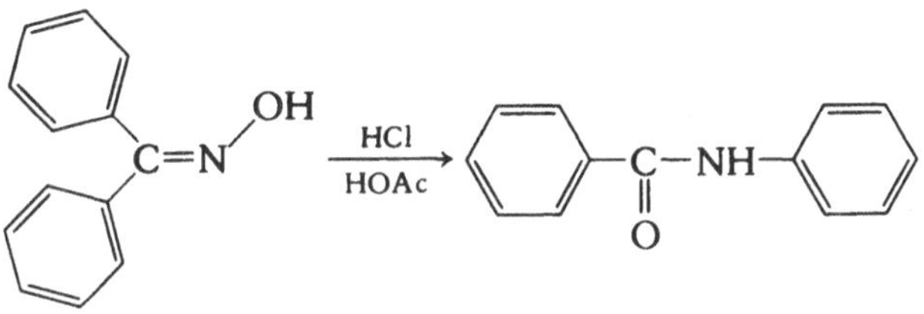

The first step in this reaction, called the Beckmann rearrangement, is protonation of the oxime oxygen. Cleavage of the N—O
bond is thereby

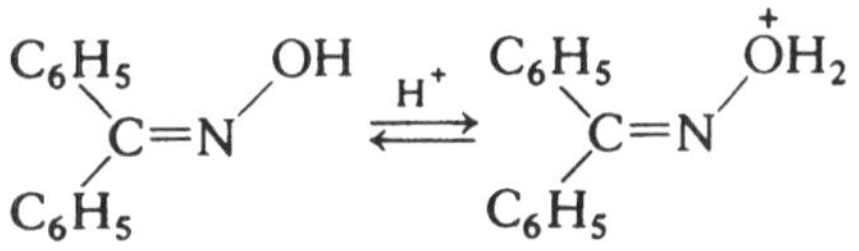

facilitated since water is a far better leaving group than hydroxide
ion. As the water departs, a phenyl group migrates to the electron-
deficient nitrogen, creating a resonance-stabilized intermediate:

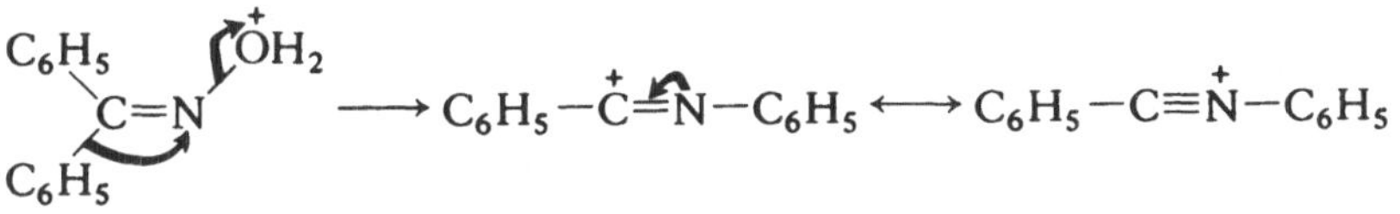

Of the two resonance contributors, the second is the more important because each atom possesses an octet of electrons whereas the
other contributor has a carbon with only six outer electrons.
Migration would occur with greater difficulty were it not for the
presence of the unshared pair of electrons on the nitrogen which
stabilizes the carbonium ion intermediate and the transition state
leading to the carbonium ion. Water (from the previous step or
from moisture in the solvent) attacks the cation; the species then

tautomerizes (much like the enol-to-ketone reaction) to the final product.

$$C_6H_5-C\equiv N-C_6H_5 \longrightarrow C_6H_5-C=N-C_6H_5$$
$$\overset{+}{H_2O} \qquad\qquad\qquad \overset{H^+}{\underset{OH}{}}$$

$$\longrightarrow C_6H_5-\underset{\underset{O}{\parallel}}{C}-\underset{\underset{}{}}{\overset{H}{N}}-C_6H_5$$

Actually, the oxime is protonated at the nitrogen to a greater extent than at the oxygen (amines are generally more basic than alcohols):

$$\underset{C_6H_5}{\overset{C_6H_5}{}}C=N\diagup^{OH} \quad\overset{H^+}{\rightleftharpoons}\quad \underset{C_6H_5}{\overset{C_6H_5}{}}C=\overset{+}{N}\underset{H}{\diagdown^{OH}}$$

However, protonation at the nitrogen is reversible and does not result in any further reaction. The small, but finite, amount of oxygen-protonated oxime in the acidic medium gives rise to the product.

25.

$$\bigcirc-CHO \xrightarrow[\text{heat}]{^-OH} \bigcirc-COO^- + \bigcirc-CH_2OH$$

Comparison of the oxidation states of the reactants and products (which is the course of action suggested before considering

any mechanism) reveals that one aldehyde molecule is oxidized
to an acid, while another is reduced to an alcohol. In the labora-
tory, carbonyls are frequently reduced to alcohols by hydride
(H^-)-donating reagents such as $LiAlH_4$, suggesting that a hydride
mechanism is operative here in the Cannizzaro reaction. Keeping
this thought in mind, we notice that hydroxide ion is an essential
entity in the reaction and that it must be utilized in the proposed
mechanism. The only reasonable reaction between hydroxide and
benzaldehyde is addition to the carbonyl group:

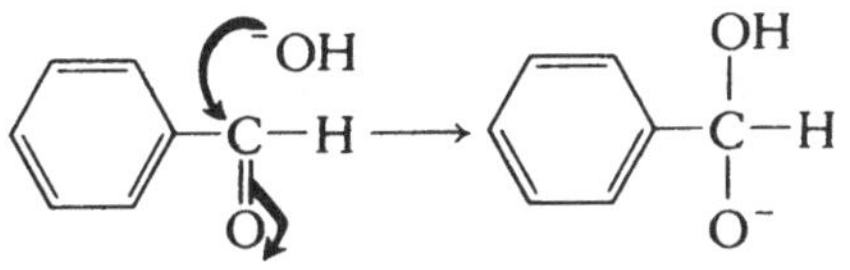

Aldehydes happen to have nonacidic protons, so that proton
abstraction by base would *not* occur:

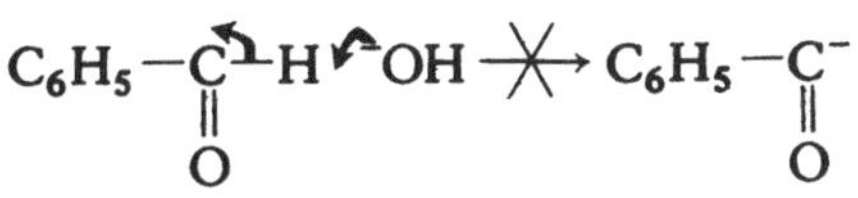

$$C_6H_5-C-H\ OH \nrightarrow C_6H_5-C^-$$

In sufficiently strong base the addition intermediate given above
can lose a proton to form a dianionic intermediate:

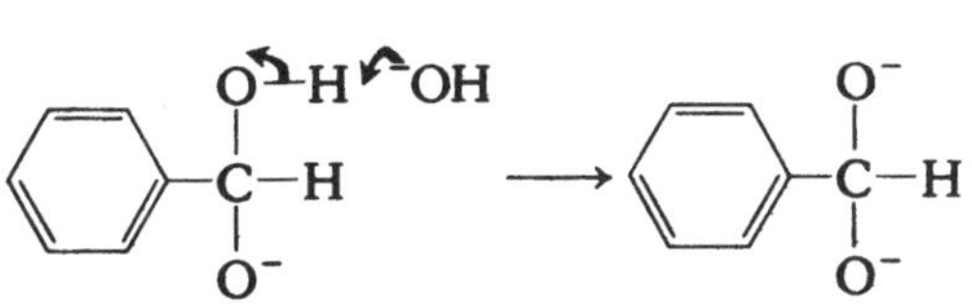

The fact that the Cannizzaro reaction is second-order in hydroxide
ion lends credence to this step. The dianion can then collapse to
a resonance-stabilized carboxylate ion by ejecting a hydride ion; a

second benzaldehyde serves as the hydride recipient:

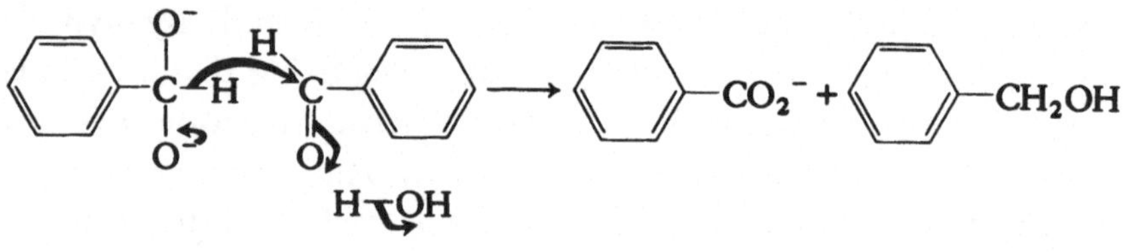

Hydride expulsion is not a particularly facile reaction; certainly the electrostatic interactions in the dianion assist the last step of the mechanism by elevating the energy of the ground state. Resonance stabilization in the benzoate ion also provides a driving force.

26.

$$CH_2{=}\overset{\overset{\displaystyle CH_3}{|}}{C}{-}CH_2OH \xrightarrow{H^+} CH_3{-}\overset{\overset{\displaystyle CH_3}{|}}{CH}{-}CHO$$

Two possible mechanisms are presented here, both of which involve formation of a carbonium ion intermediate via protonation of the methylene carbon. (We could also reversibly protonate the alcohol oxygen, but it is not obvious how the resulting species could lead to product. Protonation at the central carbon would form an unstable primary carbonium ion.) In the first mechanism, the carbonium ion is converted directly to product by means of a hydride shift:

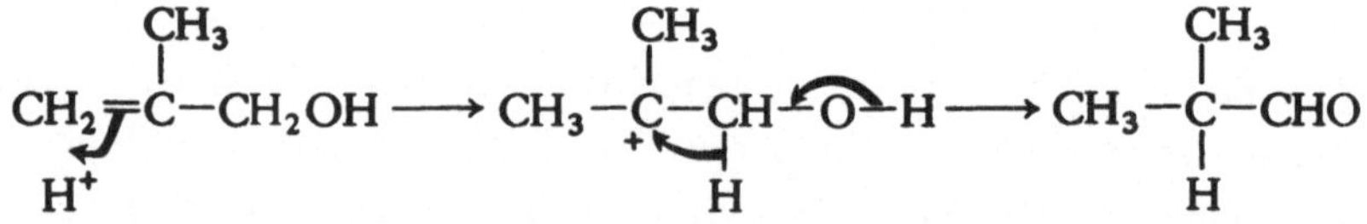

A hydrogen with a pair of electrons moves to the electron-deficient carbon and, as such, resembles a rearrangement process. In the second mechanism, loss of a proton creates a new double bond. Thus the role of the acid is to catalyze a double-bond migration. In the final step the enol rapidly tautomerizes to the more stable aldehyde form:

$$
\begin{array}{c}
\underset{\underset{H^+}{\displaystyle}}{CH_2}{=}C{-}CH_2OH \longrightarrow CH_3{-}\underset{\underset{H}{\displaystyle}}{\overset{+}{C}}{-}CH{-}OH \\
\end{array}
$$

$$
\longrightarrow CH_3{-}\underset{H^+}{C}{=}CH{-}O{-}H \longrightarrow CH_3{-}\underset{H}{\overset{CH_3}{C}}{-}CHO
$$

How might the two mechanisms be distinguished from each other experimentally?

27.

The reactant is susceptible to nucleophilic attack at three centers (namely, the two carbonyl carbons and the β carbon) all of which bear a partial positive charge. However, methoxide addition at the conjugated carbonyl and Michael addition at the β carbon are reversible and proceed no further:

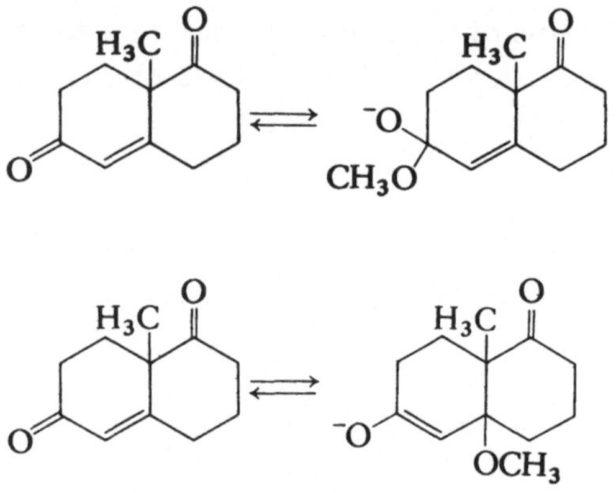

On the other hand, addition of methoxide at the third center can lead to fragmentation of the ring:

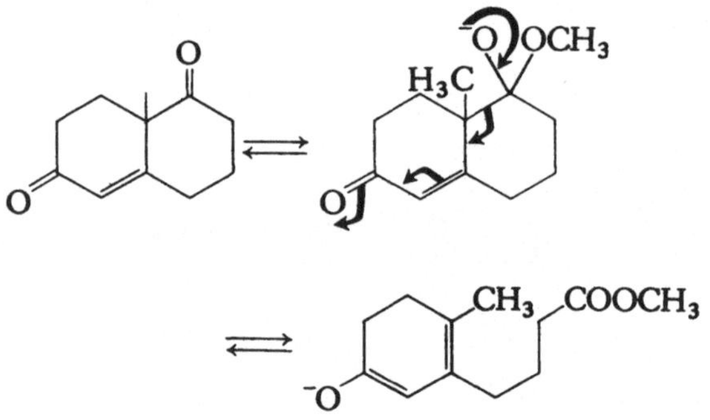

Fragmentation occurs only because a carbonyl group is positioned correctly to accept the negative charge. In the absence of the carbonyl, fragmentation would result in an unlikely, high-energy carbanion:

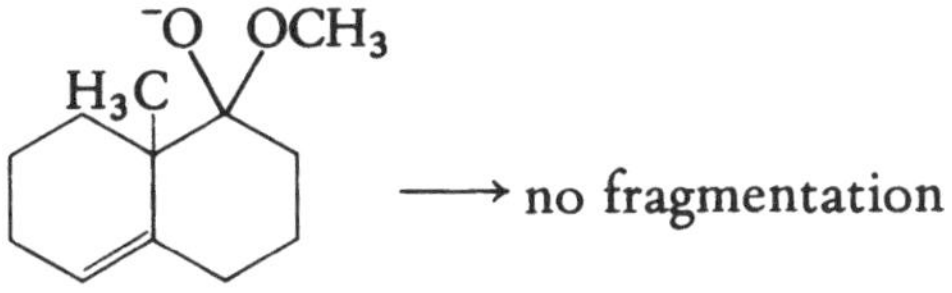

→ no fragmentation

The fragmentation product can be redrawn in another conformation:

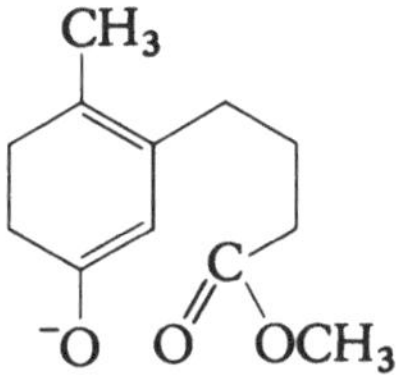

The two conformations are interconvertable merely by rotating atoms about single bonds (a very rapid process at normal temperatures). Cyclization then proceeds by means of an intramolecular nucleophilic attack on the ester carbonyl by the enolate anion:

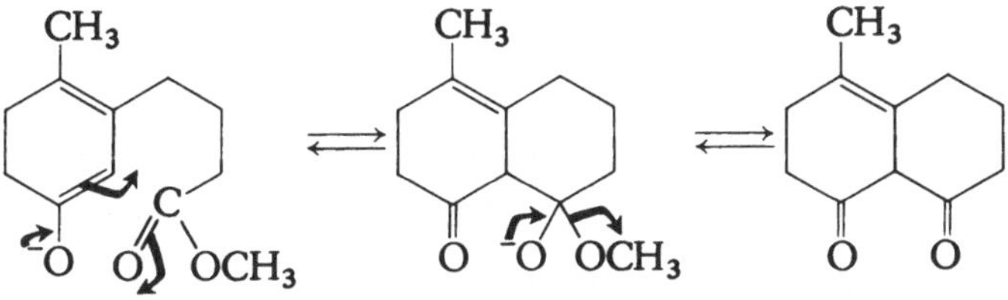

Note that all steps thus far are reversible. Since the acidity of the 1,3-diketone is similar to that of phenol, the product loses a proton in the strong base:

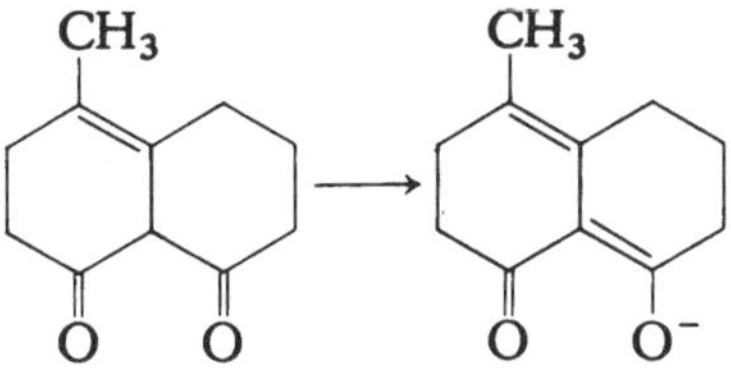

The starting material has no such acidic proton and thus the essentially irreversible formation of the anion drives the overall reaction forward.

28.

$$\text{--- } \langle \bigcirc \rangle \text{--CN} + Br(CH_2)_5 Br + POBr_3$$

The first step in analyzing a complex reaction of this sort is to compare the structures of the starting material and the products. Such a comparison reveals that:

(a) Two N—C bonds are cleaved.
(b) The carbonyl carbon loses its oxygen.

The second observation demands that the carbonyl oxygen be transformed into some functional group that can depart from the carbon. Hence, we propose that the initial step in the mechanism is nucleophilic displacement of a bromide ion from PBr_5 by the carbonyl oxygen:

Now one could argue that this is not a likely reaction because an amide oxygen is not a very good nucleophile. Even at elevated temperatures an amide does not displace bromide ion from an alkyl bromide:

$$RCONH_2 + RBr \longrightarrow \text{no reaction}$$

On the other side of the coin, PBr_5 is such a reactive species that even a poor nucleophile should be able to displace one of its bromines. In addition, the phosphorus–oxygen bond happens to be very stable. With this rationalization, we proceed to the next step in which one of the two N—C bonds is cleaved in a straight-forward (albeit somewhat unusual) S_N2 displacement:

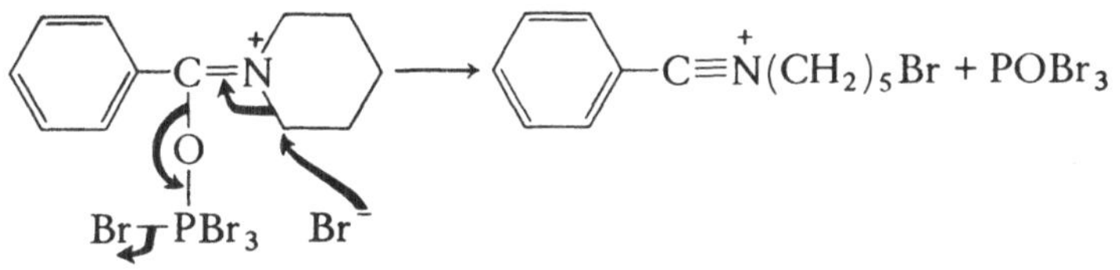

A second S_N2 displacement gives the final products:

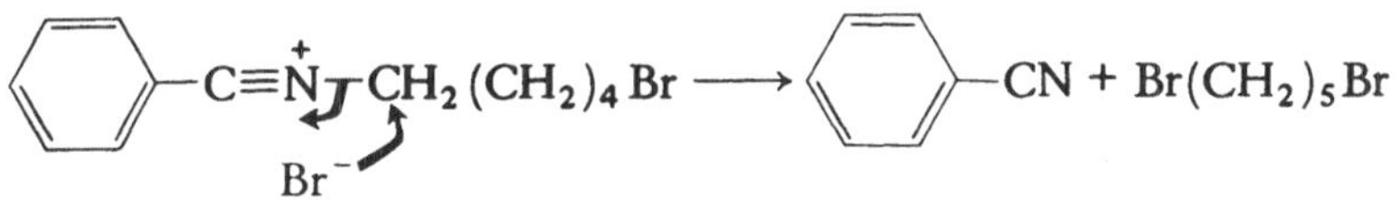

One would presume that the nitrile is an excellent leaving group in this S_N2 reaction because the nitrogen rids itself of the positive charge. Judging from the weak basicity of nitriles, the nitrile nitrogen does not readily bear a positive charge.

29.

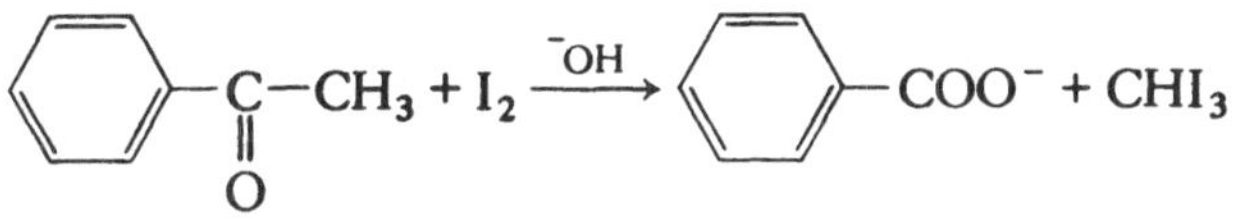

The ketone is in equilibrium with a small amount of enolate which then displaces iodide from I_2 in an S_N2-type reaction. The process repeats itself until the methyl group is totally halogenated:

$$C_6H_5-\underset{\underset{O}{\|}}{C}-CH_2-H \quad {}^-OH \rightleftharpoons C_6H_5-\underset{\underset{O}{|}}{C}=CH_2 \quad I-I$$

$$\longrightarrow C_6H_5-\underset{\underset{O}{\|}}{C}-CH_2I \longrightarrow \longrightarrow C_6H_5-\underset{\underset{O}{\|}}{C}-CI_3$$

Nucleophilic attack by hydroxide ion produces an addition intermediate which can either revert back to the components or else cleave with the expulsion of CI_3^-:

$$C_6H_5-\underset{\underset{O}{\|}}{C}-CI_3 \quad {}^-OH \rightleftharpoons C_6H_5-\underset{\underset{O}{|}}{\overset{\overset{OH}{|}}{C}}-CI_3$$

$$\longrightarrow C_6H_5COOH + {}^-CI_3 \longrightarrow C_6H_5COO^- + CHI_3$$

Carbon–carbon cleavage is made possible by the presence of *three* electronegative iodines which help stabilize the developing carbanion. The iodines reduce the negative charge on the carbon by bearing some of it themselves via inductive withdrawal:

$$\overset{\overset{I^{-\delta}}{|}}{\underset{\underset{I^{-\delta}}{|}}{{}^{-\delta}C}}-I^{-\delta}$$

Resonance also plays a role in stabilizing the carbanion; the electron pair on carbon can be donated to vacant *d*-orbitals of the 3-shell of the iodines:

$$\ddot{\overset{..}{\underset{|}{I}}}: \qquad I^{-}$$

$$\overset{\nearrow}{\underset{\underset{I}{|}}{C}}-I \longleftrightarrow \overset{\parallel}{\underset{\underset{I}{|}}{C}}-I$$

Note that the resonance contributor has an iodine with 10 electrons around it. This situation is possible only with large atoms whose bonding electrons are in the 3-shell where there are empty d-orbitals. Overlap with empty d-orbitals can occur in sulfur and phosphorus compounds as well:

$$CH_3-\overset{..}{\underset{+}{\overset{|}{S}}}-\overset{\curvearrowleft}{CH_2} \longleftrightarrow CH_3-\overset{..}{\underset{\underset{CH_3}{|}}{S}}=CH_2$$
$$\underset{CH_3}{}$$

The ejection of the carbanion in the iodoform reaction is undoubtedly also subject to steric acceleration. Both relevant carbons in the intermediate are fully substituted with bulky groups. When the carbon–carbon bond cleaves, the strain associated with the "nonbonded" interactions between the groups is relieved.

Thus, the carbon–carbon bond cleavage can be readily explained in terms of (1) inductive effects, (2) resonance effects, and (3) steric effects. These are the three most important factors used to interpret relative reactivities of closely related compounds.

30.

$$\underset{\underset{CH_2}{|}}{\overset{\overset{CH_3}{|}}{CH_3-N^{+}-CH_3}} \quad \xrightarrow[\text{liquid } NH_3]{\text{NaNH}_2 \text{ in}} \quad$$

CH_3

$CH_2N(CH_3)_2$

Proton removal from the quaternary salt produces a species known as an ylide:

$$CH_3-\overset{\overset{\displaystyle CH_3}{|}}{\underset{\underset{\displaystyle C_6H_5}{|}}{N^+}}-CH_2 \overset{\frown}{-}H \quad {}^{\curvearrowleft}NH_2 \qquad CH_3-\overset{\overset{\displaystyle CH_3}{|}}{\underset{\underset{\displaystyle C_6H_5}{|}}{N^+}}-CH_2^- \qquad \longrightarrow$$

The carbanion is not in conjugation with a carbonyl or any unsaturated group; the only stabilizing factor is the neighboring positive charge on the nitrogen which helps reduce the negative charge on the carbon by inductive withdrawal. It is no surprise that a strong base such as NH_2^- is required in order to remove protons from even a small fraction of the molecules. Note that the methylene hydrogens are somewhat more acidic than the methyl-group hydrogens because the benzyl carbanion is resonance-stabilized:

$$CH_3-\overset{\overset{\displaystyle CH_3}{|}}{\underset{|}{N^+}}-CH_3 \qquad CH_3-\overset{\overset{\displaystyle CH_3}{|}}{\underset{|}{N^+}}-CH_3 \qquad \longleftrightarrow$$

However, removal of a methylene proton does not lead to any further reaction (attack by the anionic *ortho* carbon on a methyl group would constitute a forbidden front-side S_N2 displacement).

In the next step the carbanion attacks the aromatic ring:

This process seems *unlikely* for two reasons:

(a) An electron-rich carbanion and an electron-rich aromatic ring would not have a particular affinity for one another. Aromatic rings are usually subject to *electrophilic* attack (by reagents such as H^+, Br^+, and NO_2^+) rather than to nucleophilic attack.

(b) The aromaticity of the ring and the resonance stabilization associated with it are destroyed when the intermediate is formed. Although this is not a prohibitive feature, it would certainly contribute to the activation energy of the process.

The proposed reaction must, then, be defended in the face of these objections. Two points may be made in support of the suggested mechanism:

(a) The reaction is intramolecular. In general, intramolecular reactions proceed much faster than their intermolecular counterparts when five- and six-membered transitions states are involved. For example, attack by the carbanion of the ylide at the *ortho* position of *another* ylide molecule would be out of the question.

(b) The carbanion is highly reactive. Therefore it might be expected to attack a center normally inert to more stable anions (OH^-, Br^-, CH_3O^-, etc.).

In the final step, aromaticity is secured again by means of a double-bond migration:

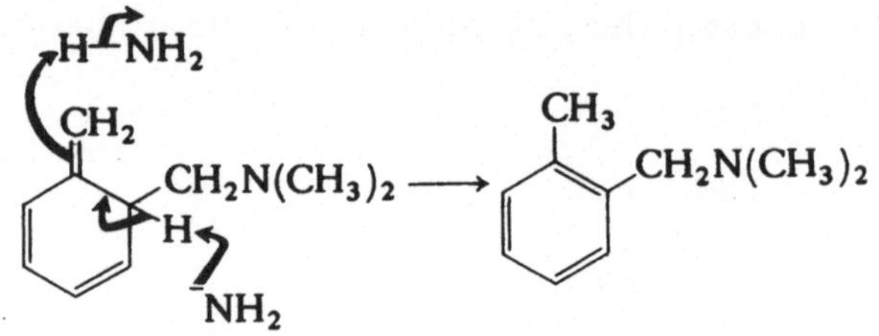

---

31. The following is a series of nonexistent, and in some cases absurd, reactions. Many of them represent errors commonly made by those with little experience in mechanism writing. Underneath each reaction is an explanation of why the reaction cannot proceed as shown.

(a)

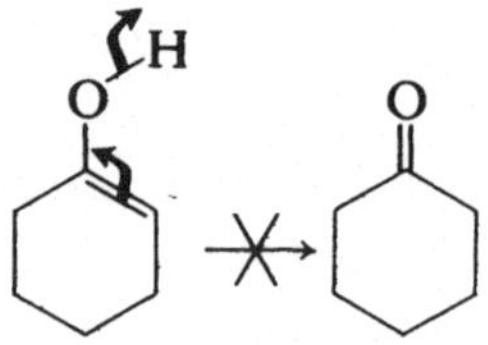

This mechanism, ostensibly an enol-to-ketone tautomerism, kicks out the hydride ion and creates a carbonium ion adjacent to a carbonyl group. The common practice of omitting protons in organic molecules tends to obscure the fact that the α carbon is cationic and bonded to only three atoms:

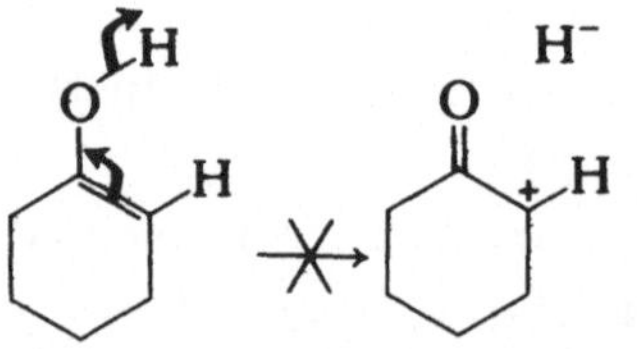

Hydride ejection cannot, of course, take place in the absence of an oxidizing agent (electron acceptor). The carbonium-ion stability would be adversely affected by the partial positive charge on the carbonyl carbon.

(b) $CH_3-C-CH_2-Br \xrightarrow{\;\;} CH_3-C=CH_2 + Br^-$

$\qquad\qquad\quad \underset{HO^-}{\overset{\longrightarrow O}{\;}} \qquad\qquad\qquad \underset{HO-O}{}$

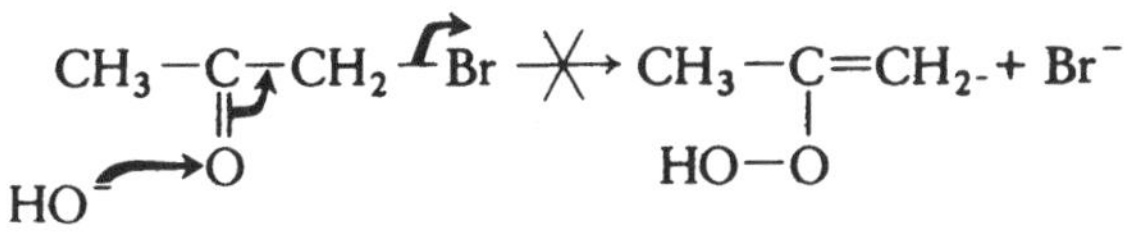

Hydroxide ion and other nucleophiles do not bond to the partially negative carbonyl oxygen. In the above "reaction," a highly energetic peroxide bond is formed without any compensating "driving force."

(c)

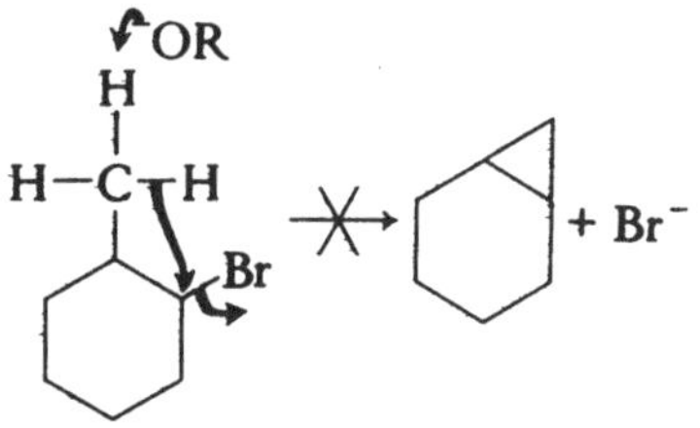

Elimination of HBr to form an alkene is, of course, a common reaction:

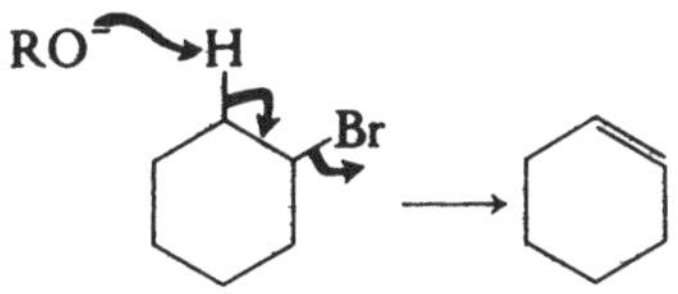

However, when the proton is further away than that shown in the elimination reaction, it is not sufficiently activated to be removed.

(d) $RCOOH + RNH_2 \xrightarrow{\;\;} R-\overset{\displaystyle O}{\underset{\displaystyle \;}{C}}-NHR$

$\qquad\qquad\qquad\qquad\qquad\qquad \underset{O}{\overset{\parallel}{}}$

When an acid and base are mixed, a *salt* is formed in an extremely fast proton transfer:

$$R-\underset{\underset{O}{\parallel}}{C}-O\text{--}H \quad R\overset{}{N}H_2 \longrightarrow R-\underset{\underset{O}{\parallel}}{C}-O^- \quad H_3\overset{+}{N}R$$

The resulting carboxylate ion is highly resonance-stabilized and inert to nucleophilic attack by all but the most reactive reagents. In order to obtain the amide, one first has to "activate" the carboxyl (e.g., convert it into an acid chloride).

(e)

The addition intermediate will not collapse to eject a totally unstabilized carbanion.

(f)
$$CH_3-\underset{}{C}=CH_2 \xrightarrow[H_2O]{H_2SO_4} CH_3-\underset{\underset{\underset{}{\smile}OH}{|}}{\overset{\overset{CH_3}{|}}{C}}-CH_3 \xrightarrow{\times} CH_3-\underset{\underset{OH}{|}}{\overset{\overset{CH_3}{|}}{C}}-CH_3$$

It is unlikely, under the acidic conditions necessary to generate the carbonium ion, that the hydroxide ion would have an appreciable concentration. Water would be a much more probable nucleophile.

(g)
$$CH_3-\underset{\underset{OH}{|}}{\overset{\overset{CH_3}{|}}{C}}-CH_3 + H_2CrO_4 \xrightarrow{\times} CH_3-\underset{\underset{O}{\parallel}}{\overset{\overset{CH_3}{|}}{C}}-CH_3$$

Avoid pentavalent carbons; it is easy to overlook this rule in a
more complex step where hydrogen atoms are omitted:

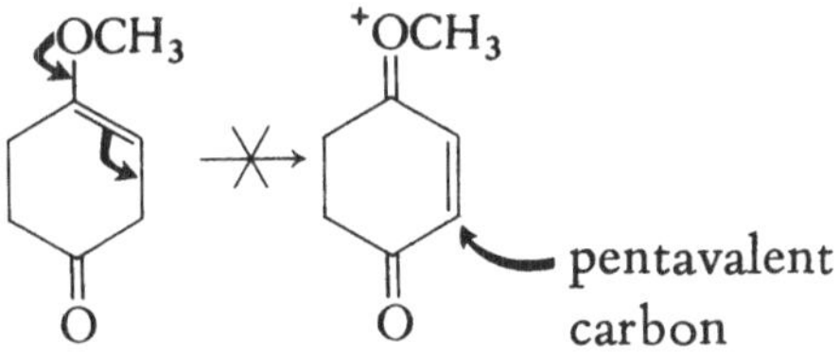

pentavalent
carbon

The charge imbalance in the above equation (neutral ⟶ positive)
also gives an indication that something is amiss.

(h)

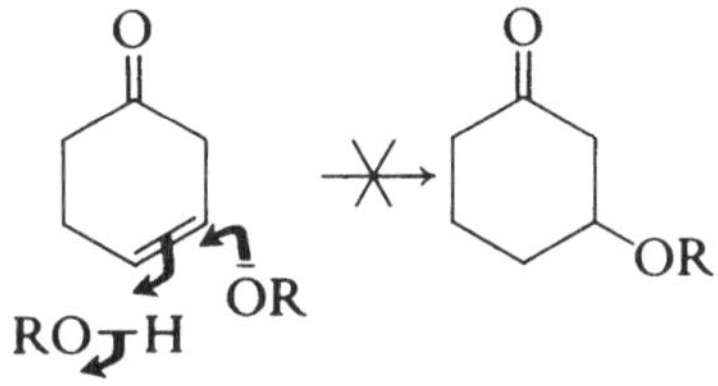

Nucleophiles do *not* attack isolated double bonds. A much more
reasonable mechanism is shown below:

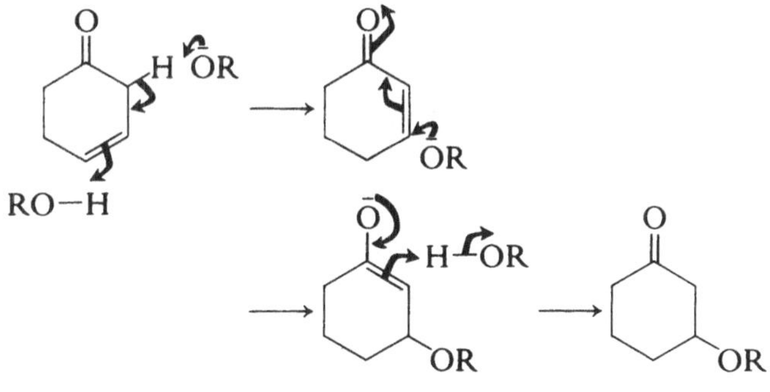

32.

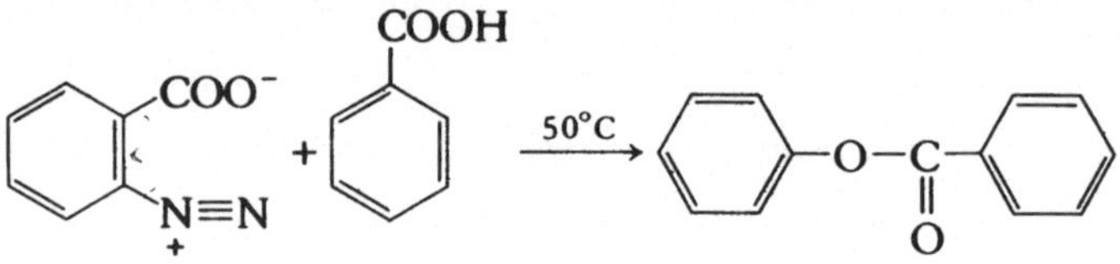

There are many remarkable organic reactions which are best rationalized in terms of a high-energy transient intermediate, benzyne. This is an example of one of them:

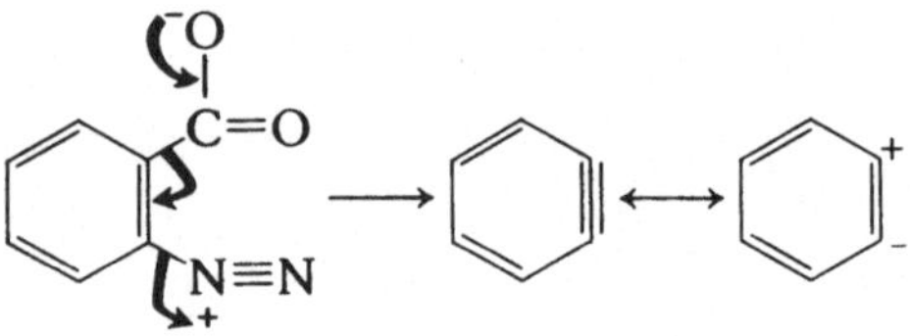

Benzyne is formed here with surprising ease; warming to 50°C is sufficient in order to eliminate CO_2 and N_2. Why should such a bizarre intermediate form so readily? Undoubtedly, two factors come into play: (a) The decomposition is associated with a favorable entropy change (one molecule ⟶ three molecules). (b) Production of two stable gases (especially N_2) serves as a "driving force." In the absence of any suitable reactant, benzyne will rapidly dimerize to biphenylene:

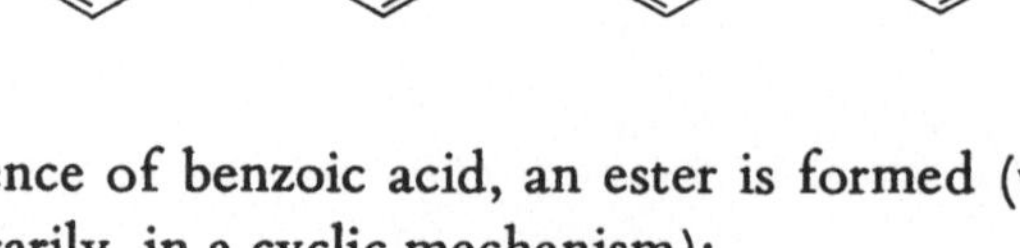

In the presence of benzoic acid, an ester is formed (written here, rather arbitrarily, in a cyclic mechanism):

Benzyne can also be formed, among other methods, by elimi-nating HX from an aryl halide in very strong base:

33.

This problem introduces another important organic inter-mediate, the carbene. Dichlorocarbene is generated from chloro-form by the action of a very strong base (a carbanion):

The carbene is a neutral divalent carbon with only six electrons in the outer shell. Because of this electron deficiency, carbenes are powerful electrophiles. They react, for example, with double bonds to form three-membered rings:

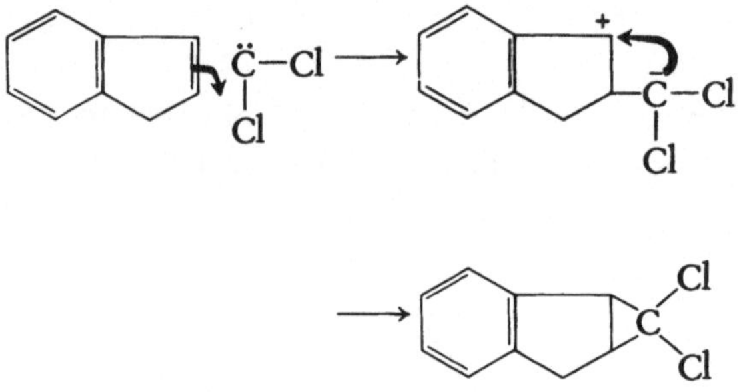

It would be a mistake to take too literally the arrows indicating electron movements; arrows are mainly electron-bookkeeping devices.

The final product is secured by an elimination reaction in which a three-membered ring is broken and a fully aromatic ring system is created:

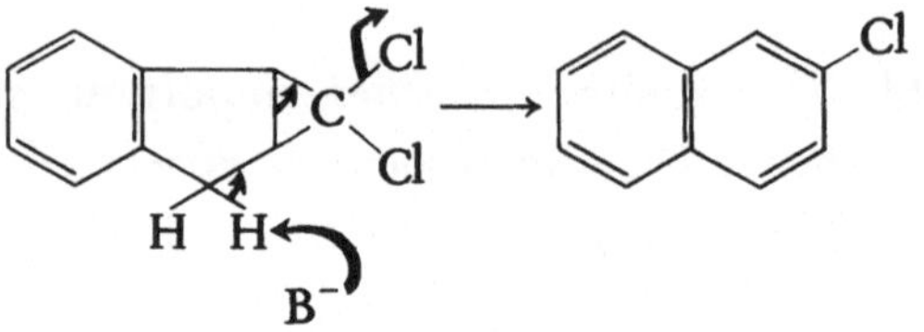

34.

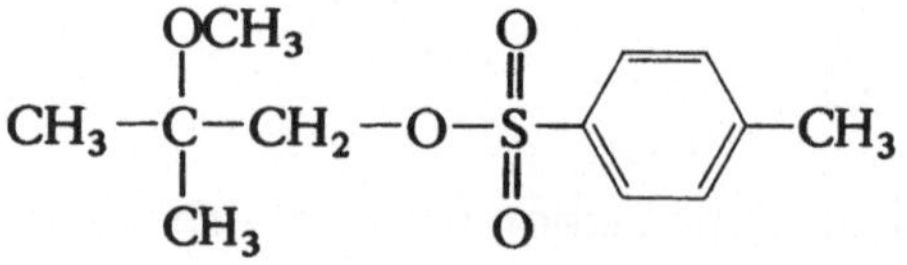

$$\xrightarrow[\text{dioxane}]{\text{aqueous}} \quad \underset{\displaystyle \overset{\displaystyle CH_3}{|}}{CH_3-CH-CHO} \; + CH_3OH + CH_3C_6H_4SO_3H$$

The reactant undergoes two particularly obvious changes in its conversion to product:

(a) The tertiary carbon loses its functional group—that is, it is reduced.

(b) The primary carbon possessing the tosylate group $(ArSO_3{}^-)$ is oxidized. The tosylate is an alcohol derivative (see below) while the product is an aldehyde.

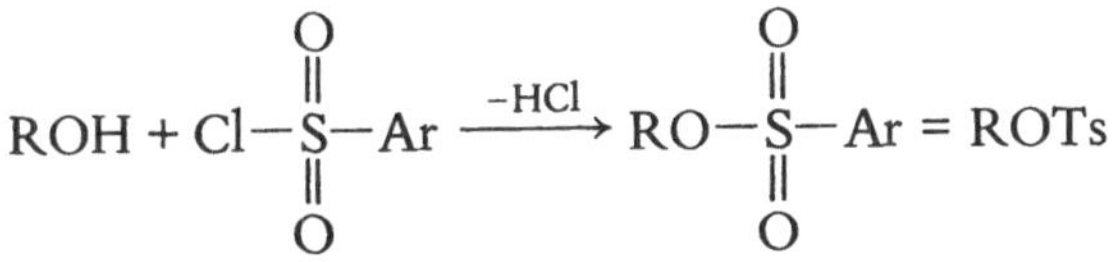

$$ROH + Cl-\overset{\displaystyle O}{\underset{\displaystyle O}{\overset{\|}{\underset{\|}{S}}}}-Ar \xrightarrow{-HCl} RO-\overset{\displaystyle O}{\underset{\displaystyle O}{\overset{\|}{\underset{\|}{S}}}}-Ar = ROTs$$

Note that the *overall* oxidation state of the reactant does not change.

In considering mechanisms for the aldehyde formation, it may be observed that if the methoxy group could somehow be interchanged with a hydrogen of the tosylate-bearing carbon, then all atoms would be in the same oxidation state as the product. (An aldehyde carbonyl carbon has the same oxidation state as a carbon single-bonded to two oxygens.) The general plan, then, is to switch groups. This can be done in a reasonable sequence of steps shown below:

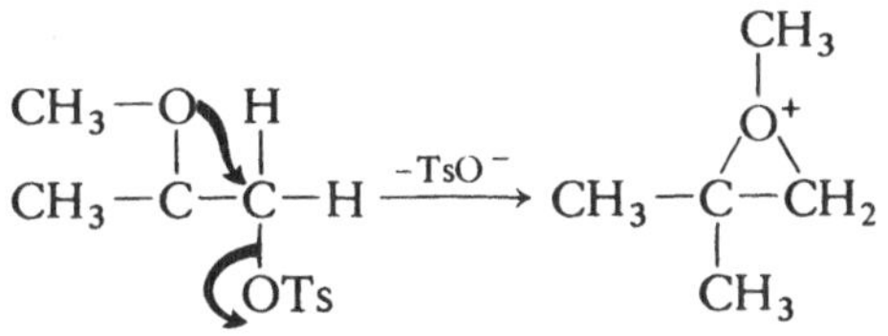

$$\underset{\displaystyle \underset{OTs}{|}}{\overset{\displaystyle CH_3-O \quad H}{\underset{\displaystyle CH_3-C-C-H}{| \quad |}}} \xrightarrow{-TsO^-} \underset{\displaystyle \underset{CH_3}{|}}{\overset{\displaystyle \overset{CH_3}{|} \;\; \overset{O^+}{\diagdown}}{CH_3-C-CH_2}}$$

The first key step is an intramolecular displacement of the tosylate by the oxygen of the methoxy group. An ether group is no doubt a poor nucleophile; on the other hand, tosylate is one of the best leaving groups known, and the reaction is intramolecular. The oxonium ion is in equilibrium with the tertiary carbonium ion:

$$CH_3-\underset{\underset{CH_3}{|}}{\overset{\overset{CH_3}{|}}{\underset{}{C}}}\overset{+}{\underset{}{O}}CH_2 \rightleftharpoons CH_3-\overset{+}{\underset{\underset{CH_3}{|}}{C}}-CH_2-OCH_3$$

It is diffuclt to guess where the equilibrium between these two intermediates lies. In any event, the products are formed by a "push–pull" hydride transfer followed by a hydrolysis:

$$CH_3-\overset{+}{\underset{\underset{CH_3}{|}}{C}}-\overset{\overset{H}{|}}{CH}-OCH_3 \longrightarrow CH_3-\overset{\overset{H}{|}}{\underset{\underset{CH_3}{|}}{C}}-CH\overset{+}{=}OCH_3 + H_2O$$

$$\longrightarrow CH_3-\overset{\overset{H}{|}}{\underset{\underset{CH_3}{|}}{C}}-CH\overset{OCH_3}{\underset{\underset{O-H}{}}{}} \quad H^+$$

$$\longrightarrow CH_3-\underset{\underset{CH_3}{|}}{CH}-CHO$$

Groups near reactive centers ("neighboring groups") can often profoundly influence the rate, course, and stereochemistry of reactions. As we saw with the $-OCH_3$ above, the neighboring group need not have great inherent reactivity to exert its effect. Several examples of neighboring-group participation are shown below:

(a)

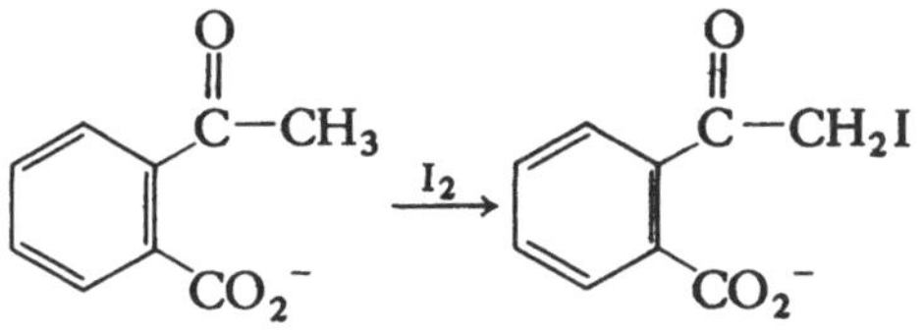

The above iodination proceeds much faster at pH 6.0 than does iodination of the corresponding ketone lacking a carboxylate. Apparently, the neighboring carboxylate assists the enolization of the ketone:

(b)

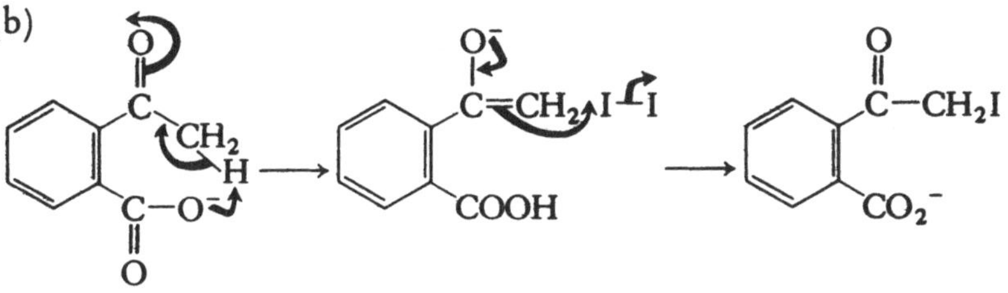

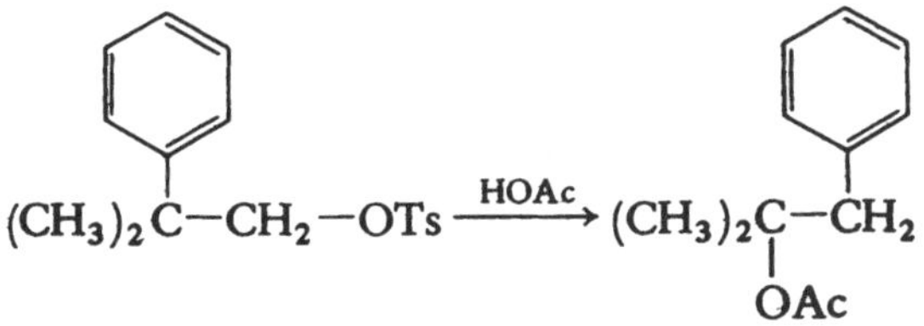

Intramolecular nucleophilic displacement leads to a rearranged product:

(c)

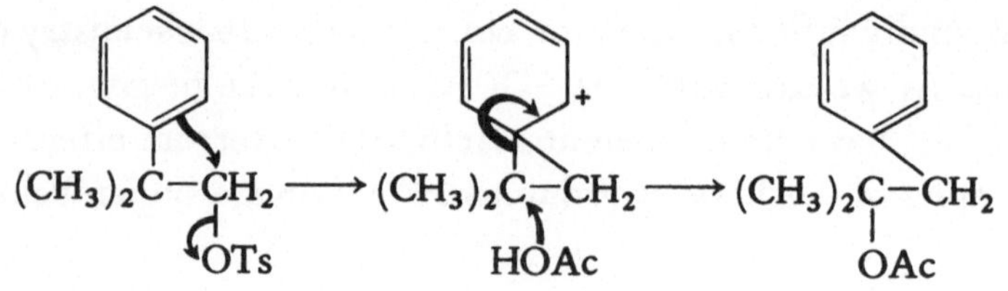

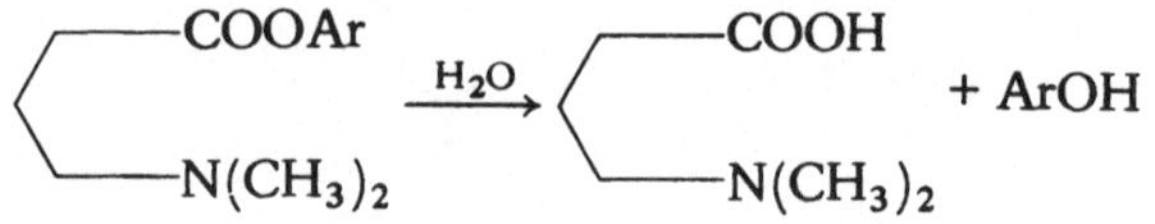

$$(CH_3)_2C-CH_2 \longrightarrow (CH_3)_2C-CH_2 \longrightarrow (CH_3)_2C-CH_2$$
$$OTs HOAc OAc$$

$$\xrightarrow{H_2O}$$ + ArOH

The tertiary amine catalyzes the hydrolysis of the aryl ester to the acid and phenol. Intramolecular nucleophilic addition to the carbonyl followed by rapid hydrolysis of the unstable acyltrialkylammonium ion gives the carboxylic acid:

(d)

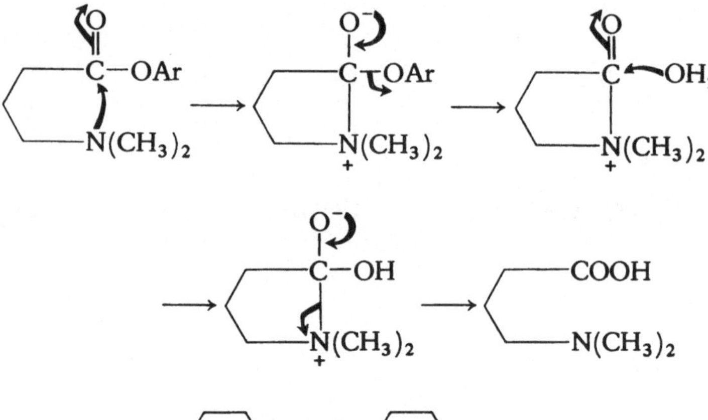

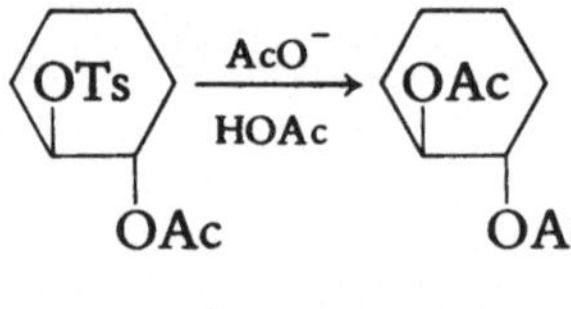

Neighboring-group participation by the acetoxy group leads to a substitution with *retention* of configuration:

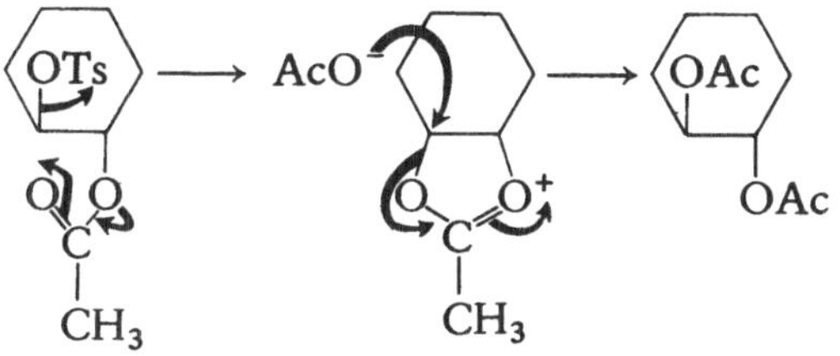

35.

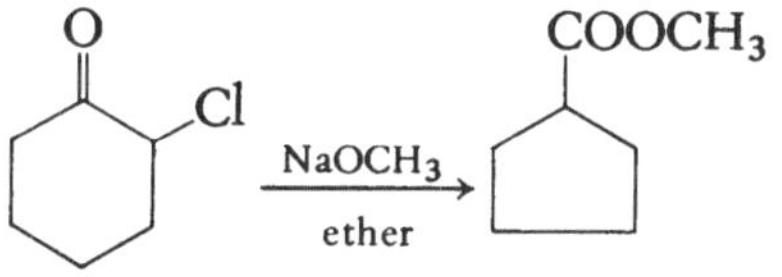

Two mechanisms have been proposed for this Favorsky rearrangement. The first involves a 1,2 migration:

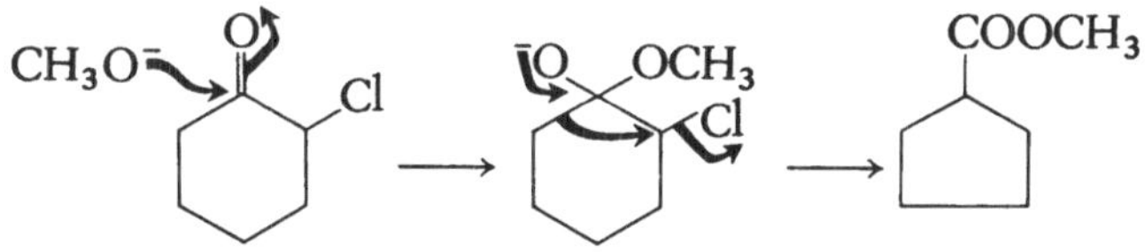

The second mechanism would appear at first sight to be less likely since it postulates the formation of a strained cyclopropanone intermediate:

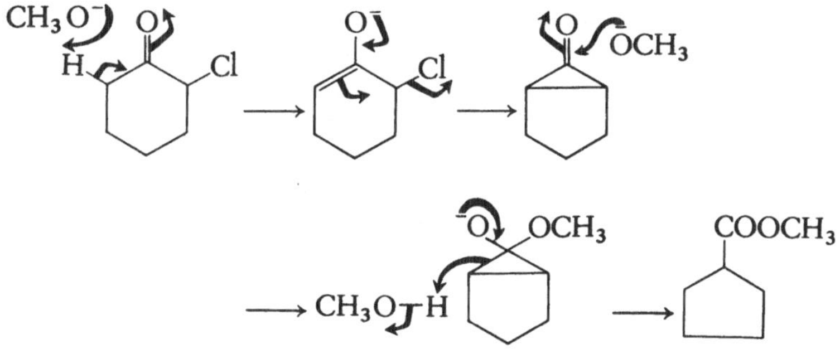

Once formed, however, the cyclopropanone would open rapidly.

It was found that 2-chlorocyclohexanone, isotopically labeled at the carbon bearing the chlorine, gives product containing labeled carbon at three positions when subjected to Favorsky conditions. The percentages of labeled carbon at the various positions are shown below:

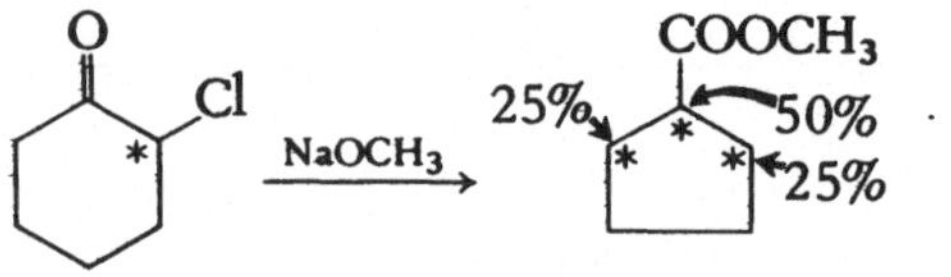

Which of the above mechanisms does this result support?

36.

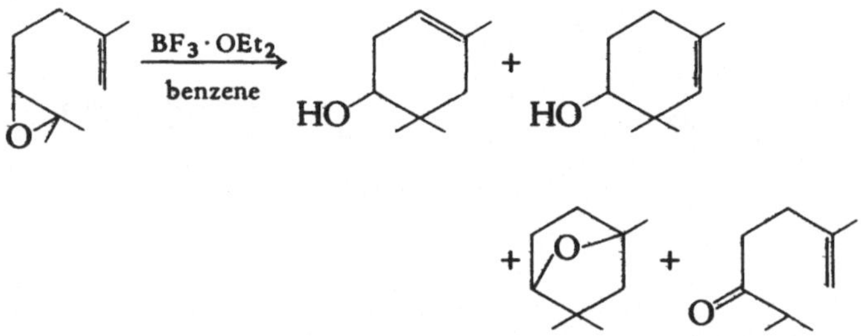

Boron trifluoride etherate is a Lewis acid and thus strongly coordinates with atoms having unshared pairs of electrons. It is often used as an acid catalyst in organic reactions because of its solubility in organic media. In the above reaction BF_3 first coordinates with the epoxide oxygen to yield an oxonium ion:

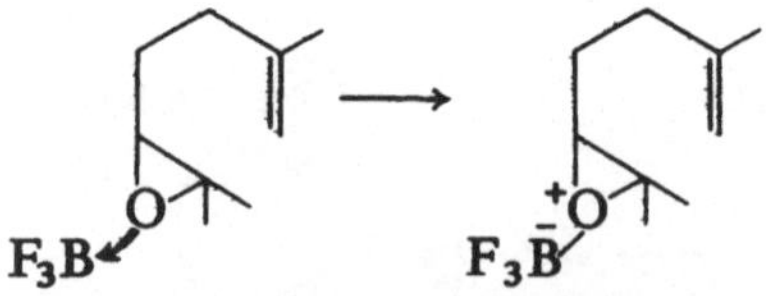

The oxonium ion may now open, relieving the strain in the three-membered ring:

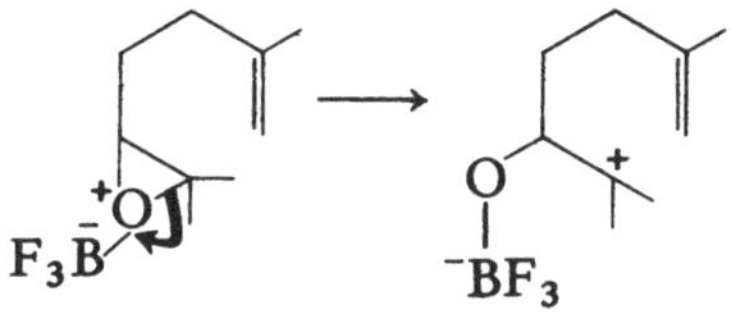

Note that the epoxide opens toward the tertiary center rather than the secondary carbon:

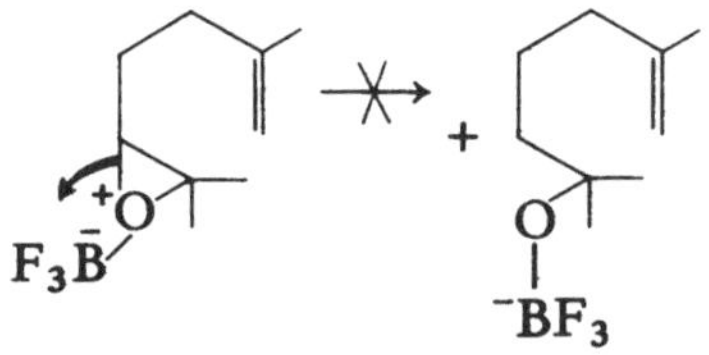

In any event, the electrons in the neighboring double bond can be used to form a new carbon–carbon σ bond thereby generating another tertiary carbonium ion:

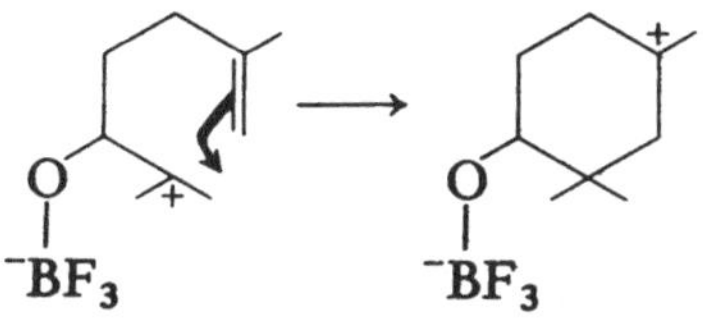

Alternatively, one could open the epoxide and form the six-membered ring in a single step.

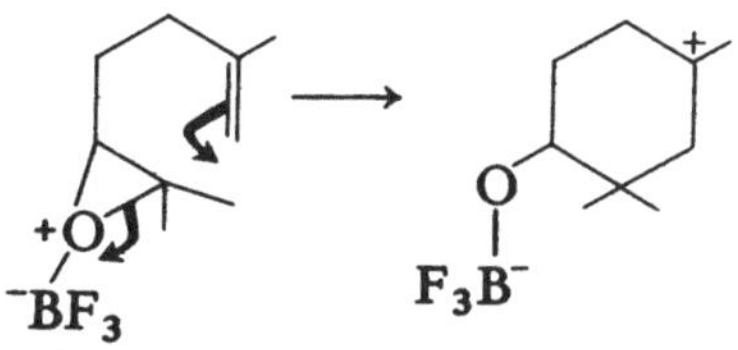

Proton loss from one or the other of the methylenes flanking the carbonium ion gives the first two of the four products (after work-up in water, which destroys the BF$_3$ derivative of the alcohol). For example:

The carbonium ion can also react intramolecularly with the oxygen:

Finally, a ketone may be generated by a hydride shift:

37.

This is an example of a *fragmentation reaction* in which the strong base removes the proton from the alcohol, followed by cleavage of a carbon–carbon bond and elimination of tosylate:

38.

This Barton reaction represents one of the few examples in which an unactivated carbon (the methyl group between rings C and D of the steroid) becomes functionalized with great regiospecificity.

The process is initiated photochemically, with the nitrite decomposing to the alkoxy radical:

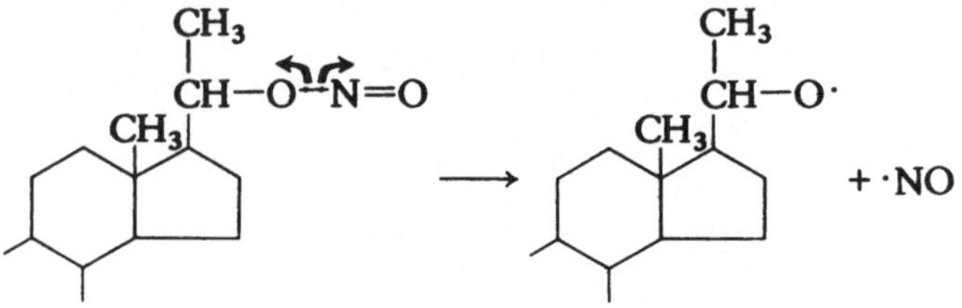

Homolytic cleavage of the oxygen–nitrogen bond is favored by the relative stability of the nitroso radical. The alkoxy radical then abstracts a hydrogen atom from the neighboring methyl group to generate a primary radical:

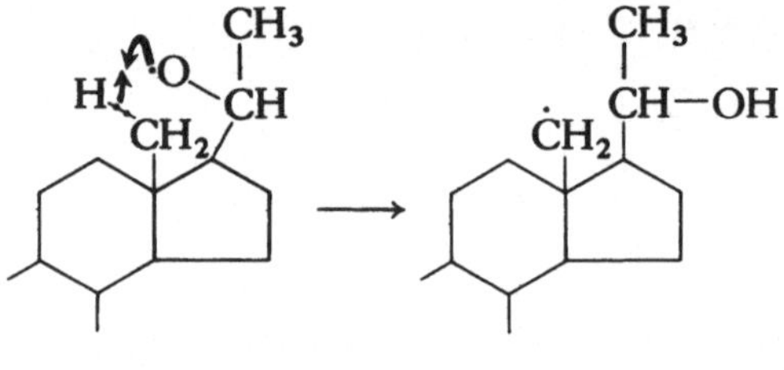

The success of this hydrogen atom transfer reflects the proximity of the alkoxy radical to the methyl group. This is perhaps better seen in a nonplanar representation:

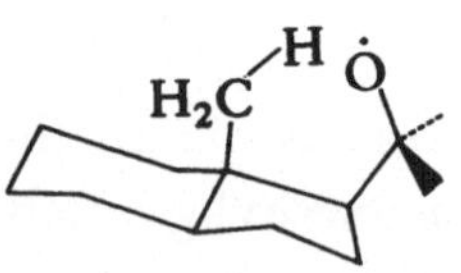

No hydrogen other than those on this particular methyl group is so favorably disposed toward the alkoxy radical. In the next step, the primary radical couples with the nitroso radical to afford a nitroso intermediate:

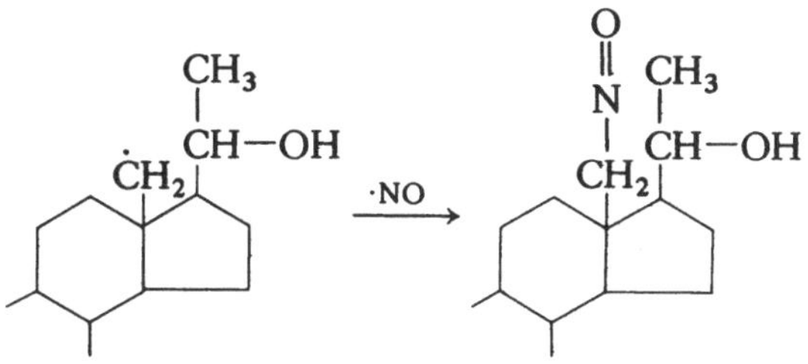

It appears that the nitroso species that is formed in the first step is not completely "freed"; that is, the nitroso radical bonds to the carbon on the *same* molecule from which it originally departed. Isomerization of the nitroso group to the oxime gives the final product. This last reaction resembles enolization of a carbonyl:

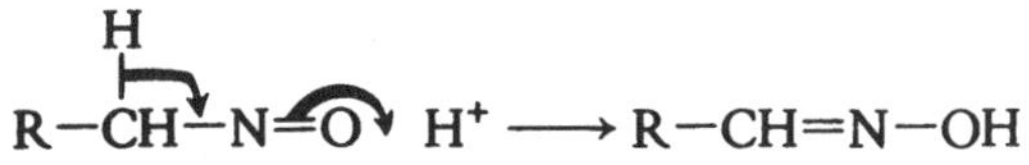

39.

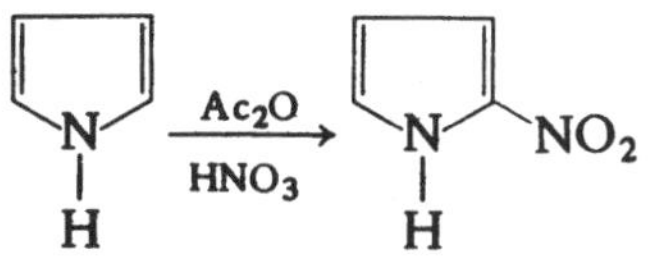

Acetic anhydride assists the formation of a potent nitrating agent, NO_2^+ (nitronium ion); the probable involvement of a tetrahedral intermediate is indicated using a "shorthand" notation:

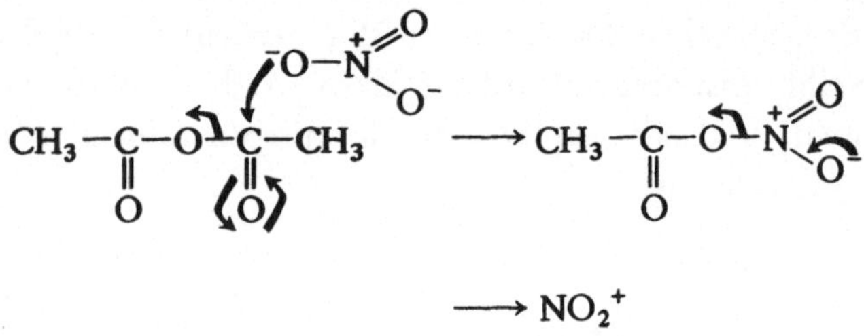

$$\longrightarrow NO_2^+$$

The electrophilic nitronium ion then attacks the electron-rich pyrrole:

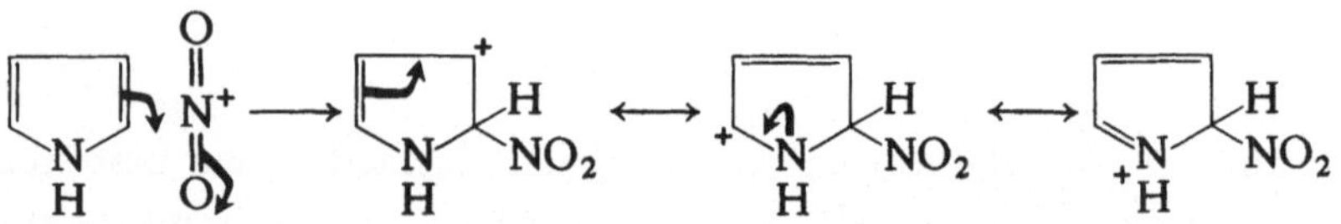

The resulting addition intermediate is highly resonance-stabilized. Loss of the proton reestablishes the pyrrole system:

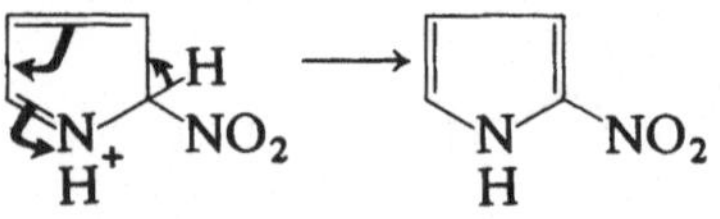

Note that addition of NO_2^+ at the alternative pyrrole carbon would form an intermediate with only two resonance contributors:

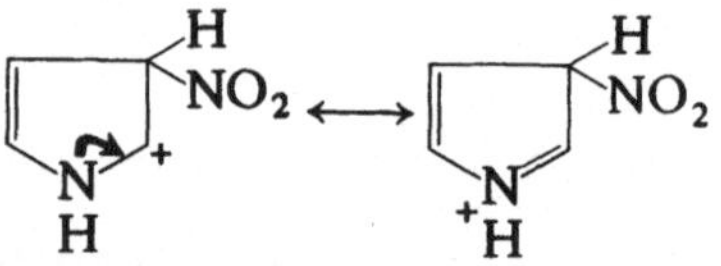

Hence, this mode of addition would not be as likely. The student should attempt to explain why chlorination of indole with Cl^+ leads to substitution at the 3-position:

40.

The transformation of the reactant, called thebaine, to the product appears to involve a drastic restructuring of the molecule. It can be relatively easily rationalized, however, in terms of carbonium-ion stability. Initially, the ether is cleaved to form an allylic-type carbonium ion:

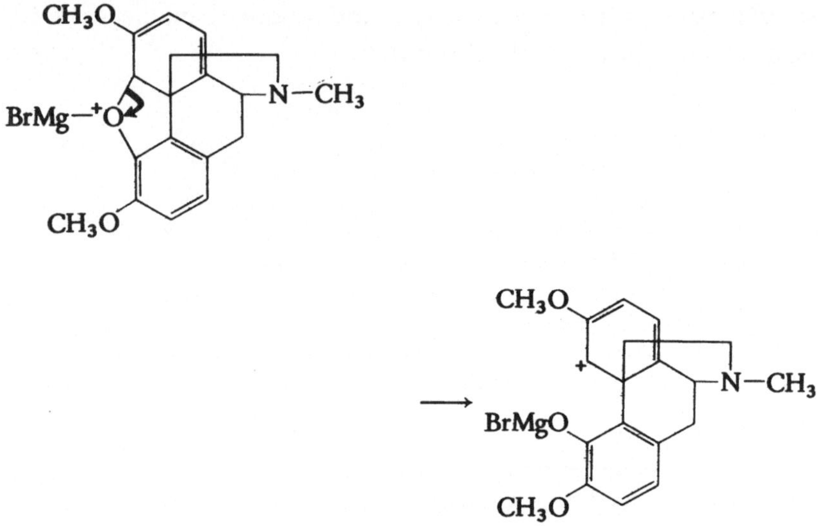

Bond breakage does not occur between the ether oxygen and the aryl ring because this would lead to an unstable aryl carbonium ion:

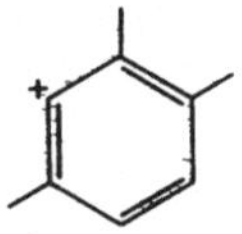

Note that the allylic carbonium ion has a resonance contributor in which the positive charge resides on a tertiary carbon:

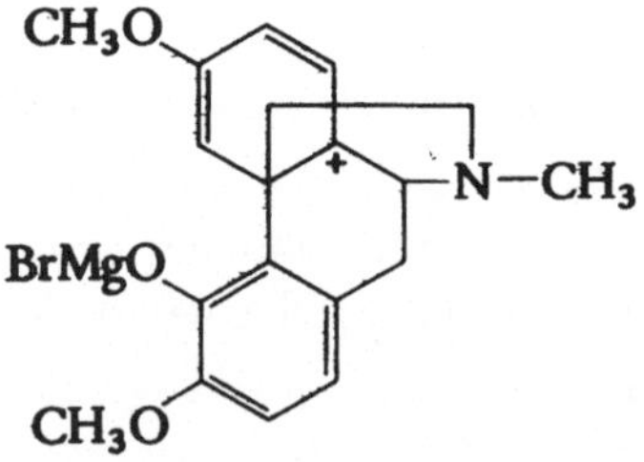

Wagner–Meerwein rearrangement of the carbon skeleton to this electron-deficient center produces a new carbonium ion:

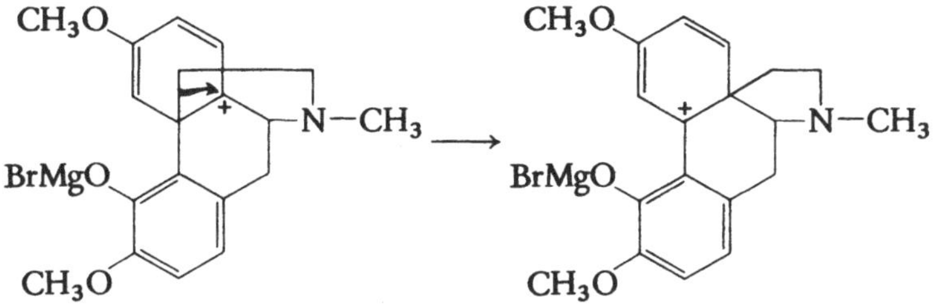

The driving force for this rearrangement is the additional resonance stabilization accompanying formation of a tertiary, allylic, and benzylic carbonium ion. The next step is also favorable from an energetic point of view:

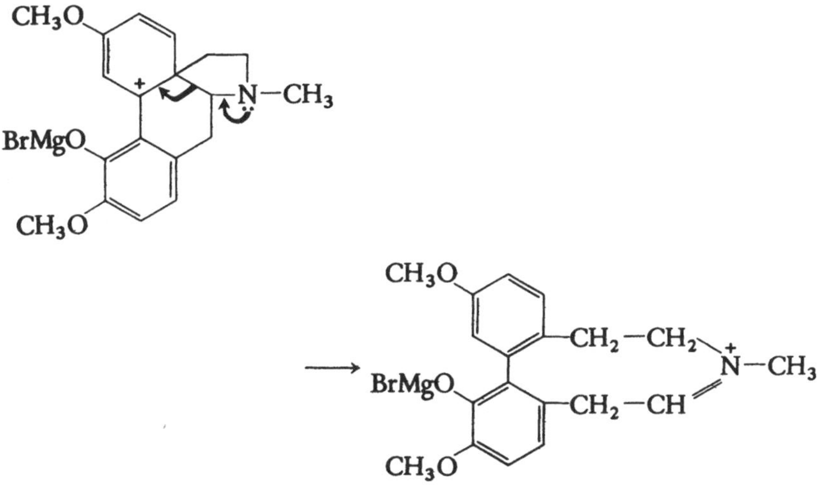

A new aromatic ring is constituted and the positive charge is placed on a nitrogen such that all atoms now possess a full octet

of electrons. In the last step the phenyl carbanion from the Grignard reagent adds to the immonium ion in accord with simple
Grignard additions:

CHAPTER 3

Problems in Organic
Reaction Mechanisms

3.1. Problems

1.

2.

3.

4.

5.

6.

7.

8.

9.

10.

$$C_6H_5S-CH_2-CH=C\overset{CH_3}{\underset{CH_3}{\big\langle}} + :CCl_2$$

11.

12.

13.

14.

15.

16.

17.

18.

19.

CH_3O (dihydropyran with OMe) $\xrightarrow[\text{CH}_3\text{CN}]{\substack{C_6H_5SO_2N_3 \\ \text{reflux in}}}$ CH_3O—(piperidinone) N—$SO_2C_6H_5$, ring C=O

20.

(4-benzoyloxycyclohexanone; ring with O at top, $O\overset{\|}{C}-C_6H_5$ at bottom) $\xrightarrow{t\text{-BuO}^-}$ $C_6H_5-\underset{\underset{O}{\|}}{C}-CH-CH(CH_2)_2COOH$ with $\underset{CH_2}{\overset{|}{}}$ forming a cyclopropane (CH_2 bridge between C and CH)

21.

(2-acetylfuran) $\underset{O}{\overset{\|}{C}}-CH_3$ $+$ (morpholine, $\overset{H}{N}$ ... O) $\xrightarrow[\text{HCl}]{\text{trace of}}$ (benzene ring with OH and N-morpholine substituents)

22.

$CH_3-\underset{\underset{O}{\|}}{C}-COO^-$ $+$ (thiazolium salt: R, S, $\overset{+}{N}$, R; R; catalyst) $\longrightarrow$ $CH_3-\underset{\underset{O}{\|}}{C}-\underset{\underset{OH}{|}}{C}H-CH_3$

(catalyst)

23.

24.

25.

26.

3.2. Answers

1.

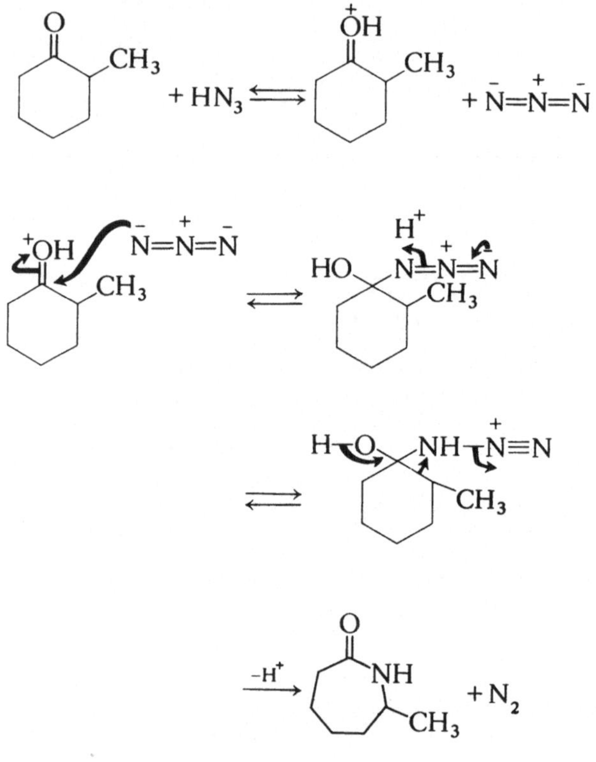

2.

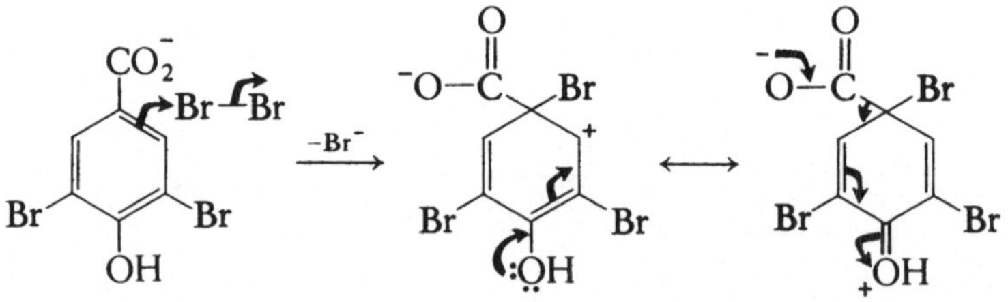

$$\xrightarrow{-\overset{\cdot}{C}O_2}$$

3.

$$\xrightarrow{CH_3MgX}\quad\xrightarrow{-CH_4}$$

$$\xrightarrow{-CN^-}$$

$$\xrightarrow[\text{2. } H_2O \text{ (in work-up)}]{\text{1. } CH_3MgX}$$

4.

Note that a direct S_N2 displacement of trimethylamine by the carbanion is very unlikely; the amine functionality is an extremely poor leaving group in substitutions. The success of this S_N1 substitution reflects the stability of the carbonium ion intermediate.

5.

6.

The final step in this reaction is an example of a "1,3 dipolar addition." Why does none of the isomer below form?

7.

8.

9.

10.

and

11.

12.

What is the driving force for the final step in the mechanism?

13.

14.

15.

16.

17.

18.

19.

Since the above mechanism entails an unstabilized secondary carbonium ion, it might be preferable to form the final intermediate directly:

20.

$$\longrightarrow C_6H_5-\underset{\underset{O}{\|}}{C}-\underset{\underset{\overset{|}{CH_2}}{}}{CH}-CH(CH_2)_2COOH$$

21.

22.

23.

24.

25.

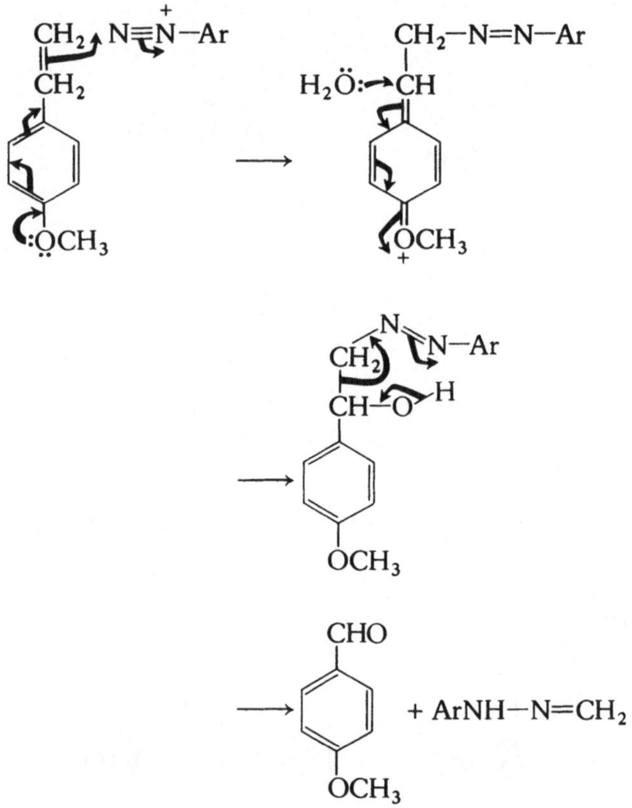

CH$_2$—N≡N$^+$—Ar
‖
CH$_2$

:OCH$_3$

$\longrightarrow$

CH$_2$—N=N—Ar
H$_2$Ö:→CH

ŌCH$_3^+$

$\longrightarrow$

N
‖
CH$_2$ N—Ar
CH—O—H

OCH$_3$

$\longrightarrow$

CHO

OCH$_3$

$+$ ArNH—N=CH$_2$

ArNHN=CH$_2$ + H$_3$O$^+$ ⇌ ArNHNH$_2$ + CH$_2$O

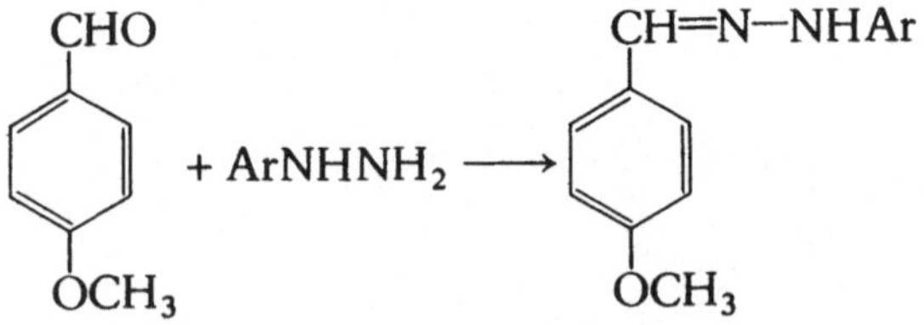

CHO

OCH$_3$

$+$ ArNHNH$_2$ $\longrightarrow$

CH=N—NHAr

OCH$_3$

26.

Transannular ether–oxygen participation (next to last step of the mechanism) could conceivably occur at an earlier stage.

CHAPTER 4

A Molecular Orbital Approach to Organic Chemistry

4.1. Basic Principles

Simplicity is one of the main attributes of "electron pushing." Using a relatively small number of basic rules, one can rationalize a vast number of organic reactions, predict the products of reactions on new systems, and design new reactions. It does not really matter if electron pushing has little physical reality. What does matter is that the method works and that it is fast and simple. Sophisticated theoretical approaches, often requiring prodigious amounts of computer time, can also predict organic chemical behavior, but they are often not useful in the day-to-day routine of most organic chemists.

Analyses of reaction mechanisms by electron pushing have important limitations. For example, it is not possible to determine by means of arrows whether S_N2 reactions should occur with front-side displacement or back-side displacement:

Front-side

Back-side

$$B \overset{\frown}{} \overset{\Rightarrow}{\underset{/}{C}} - X$$

Similarly, "arrows" give no information as to geometry of eliminations:

$$B \overset{\frown}{} H \qquad \qquad B \overset{\frown}{} H \quad X$$

or

In this chapter we will discuss molecular orbital (MO) methods
that handle such stereochemical subtleties. This is not to imply
that the MO method is in any way superior to electron pushing. They
do different things, and both techniques are needed by organic
chemists. We should also point out that our coverage of molecular
orbitals is elementary and totally nonmathematical, yet we hope
it will provide insights into organic reactivity which complement
those acquired in the first part of this book.

Formation of H_2 from two hydrogen atoms can be represented in arrow notation as follows:

$$H \cdot \overset{\frown}{} \cdot H \longrightarrow H_2$$

In MO terms, the hydrogen–hydrogen bond can be viewed as the
combination of *two* $1s$ atomic orbitals to form *two* molecular
orbitals. The number of interacting atomic orbitals always equals
the number of molecular orbitals formed. In the case of H_2, one
of the molecular orbitals is a low-energy *bonding orbital* filled
with the two electrons. The other is an empty high-energy *anti-*

bonding orbital:

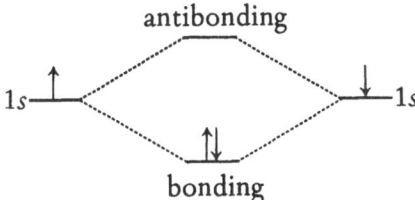

In the above diagram electrons are represented by arrows; the different direction of the two electrons in the bonding MO arises from the fact that electrons must spin in opposite directions to be accommodated in an orbital. Of course, no orbital can contain more than two electrons. The bonding or σ orbital can be crudely depicted as two overlapping spheres:

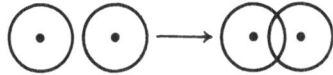

The overlap signifies that there is considerable electron density between the two positively charged nuclei. Hence the electrons "bond" the hydrogen atoms which, in the absence of the electrons, would normally electrostatically repel each other. The unfilled antibonding or σ* MO can be represented as

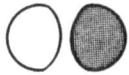

Strong internuclear repulsion is created by the absence of overlap and by the low electron density between the atoms. The components of the antibonding MO are shown in two colors, gray and white (relating to different signs in the wave function describing the orbital). For our purposes it is sufficient to accept the fact that overlap is permissible when the colors (signs) are the same and not possible when the colors (signs) are opposite. Note that in the antibonding MO there is a zero probability of finding an electron midway between the hydrogen atoms; such a point is called a node. It is easy to assess the relative energy of a series of MO's for a given system: the fewer the number of nodes, the lower the energy of the orbital. In accordance with this rule, the bonding orbital of H_2 has no node.

The methyl radical, possessing an electron in an sp^3 orbital, is shown below:

When two methyl radicals react to form ethane, the atomic sp^3 orbitals combine to form two MO's:

—— Antibonding (σ^*)

⇅ Bonding (σ)

There is no distinction here between gray and white in the sense that we could have drawn the bonding MO as

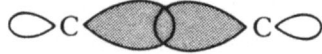

and the antibonding MO as

The important thing to note is that the bonding MO features the *same* color between the carbon atoms (indicating favorable overlap) whereas this is not true of the antibonding MO.

The relative size of the lobes provides a rough indication of the probability of finding an electron at a given location. Thus, the sp^3 atomic orbital of the methyl radical possesses only a small lobe in the "crotch" formed by the three protons; the electron will be found most often in the region outside the crotch defined by the larger lobe. Similarly, an electron in the bonding MO of the carbon–carbon bond locates itself generally between the carbon atoms; this high electron density between the carbons in the bonding MO is pictorially represented by the large pair of internal lobes. In contrast, the antibonding MO, which does not contribute to the bonding of the atoms, has its larger lobes (i.e., electron density) external to the carbon–carbon linkage.

In order to apply MO theory to simple organic reaction mechanisms, one must invoke two key rules:

(a) Bond formation occurs when an *occupied MO* interacts with an *unoccupied MO*.
(b) All interacting orbitals must match if a reaction is to occur readily. Mismatches lead to high-energy transition states.

Consider, for example, the S_N2 displacement reaction. A nucleophile N with a pair of electrons in an occupied sp^3 orbital reacts with the unoccupied antibonding MO of the carbon–halogen σ bond:

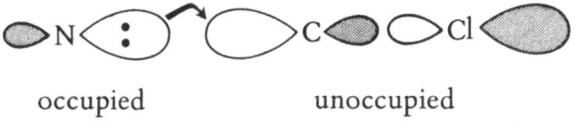

occupied   unoccupied

The arrow joins two lobes of the *same* color, and this matching leads to the conclusion that back-side S_N2 attack is permissible. Of course, we know that back-side attack does in fact occur because S_N2 reactions take place with inversion of configuration:

The occupied sp^3 orbital of the nucleophile can also be drawn as

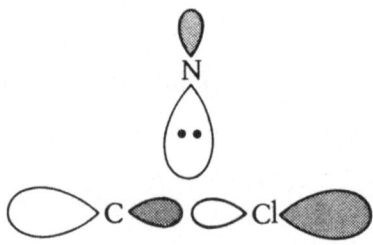

This representation would not match with the unoccupied MO as given above, but this is irrelevant; the point is that one *can* draw matching orbitals leading to bond formation. Note that back-side attack not only matches orbitals but also involves the *larger* of the two carbon lobes comprising the antibonding MO. Thus overlap is maximized in the transition state.

There is not a single proven case of front-side S_N2 displacement. In orbital terms this can be explained by a match canceled by a mismatch:

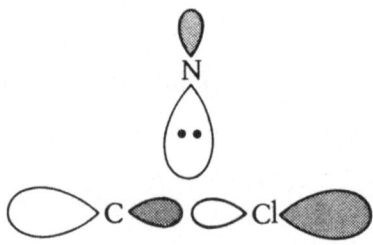

Even the match involves a *smaller* lobe in the antibonding MO of the carbon–chlorine bond. Both factors contribute to the unlikelihood of the reaction.

As with the electron pushing approach, the above MO explanation of the S_N2 stereochemistry is a gross oversimplification. This is of little concern to us because we get the right answer and can even use the method for predictive purposes, as we shall see shortly.

The overwhelming majority of bimolecular aliphatic *electrophilic* substitution reactions occur with *retention* of configuration (i.e., front-side attack). Two examples are given below:

In arrow notation we can depict this type of reaction as

Why should a nucleophilic substitution occur with inversion whereas an electrophilic substitution prefer retention? In answering this question using MO theory, we can reasonably postulate that an empty Y orbital interacts with a filled (bonding) C—X MO. The situation for a back-side attack by electrophile Y would then be

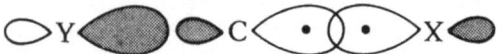

Matching is seen to be possible, and in fact a few cases of back-side electrophilic substitutions are known. Compare this, however, with the front-side reaction:

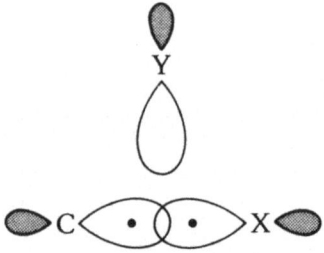

The front-side mode clearly represents a superior electronic interaction. All three lobes match and, in addition, it is the larger pair of lobes that is involved in the substitution process. Stated another way, the electrophile reacts at the site of highest electron density, which is, according to the orbital picture, located between the C and X atoms.

The elimination reaction beautifully illustrates the power of the MO approach. In a typical E2 elimination, a strong base is used to initiate removal of HX:

$$\text{HO} \frown \text{H}$$
$$R_2C \overset{\curvearrowleft}{-} CR_2 \longrightarrow R_2C=CR_2$$
$$\underset{X}{\big|}$$

Spontaneous elimination of HX without any base according to the mechanism below is, understandably, energetically unfavorable because two ionic species would be generated:

$$R_2C-CR_2 \nrightarrow R_2C=CR_2 \quad H^+ \quad X^-$$

Since these ions must be stabilized by solvation, many solvent molecules would be "frozen" (an entropically costly process). On the other hand, it is much less obvious why a cyclic concerted reaction which avoids ion production should not occur:

$$R_2C-CR_2 \nrightarrow R_2C=CR_2 + HX$$

This reaction cannot proceed readily or else organic halides would not be stable compounds. Yet, as far as "electron pushing" is concerned, there is nothing particularly wrong with the mechanism (except perhaps the feeling that a four-membered transition state leaves something to be desired). The MO description resolves the dilemma. One can regard the cyclic concerted mechanism as an interaction of electrons in the C—H occupied MO with the unoccupied MO of the C—X:

Such a cyclic mechanism is energetically unfeasible because it would yield H—X in the antibonding state owing to the mismatch. *All* sites in a concerted reaction must match properly, and violation of this basic rule in the cyclic elimination gives a high-energy transition state leading to an unacceptable product.

Having disposed of the cyclic mechanism, let us now rationalize the observation that E2 eliminations prefer a *trans*–coplanar relationship in which H, X, and the two central carbons are in the same plane, and H and X are on opposite sides of the carbon–carbon bond:

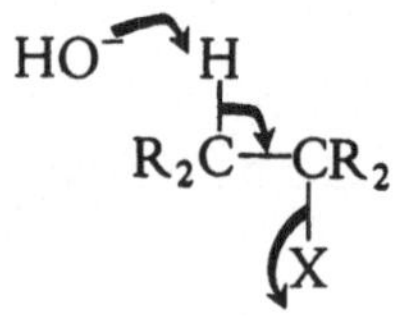

$$HO \quad H$$
$$R_2C - CR_2$$
$$X$$

For example, elimination of HBr from *meso*-1,2-dibromo-1,2-diphenylethane gives exclusively the *cis*-2-bromostilbene:

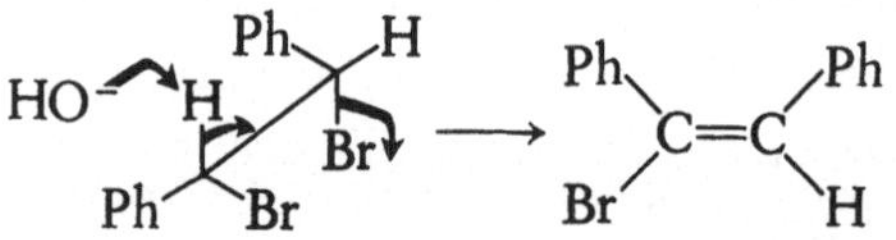

So-called *syn* elimination would have produced the *trans* product:

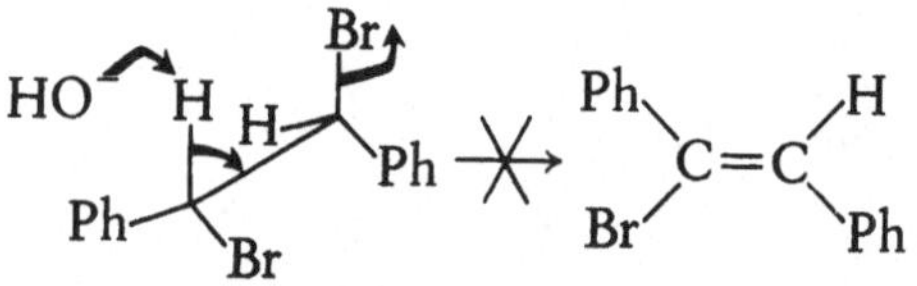

Examples of *syn* elimination are known, but they are rare and usually require more severe conditions (stronger base or higher temperature):

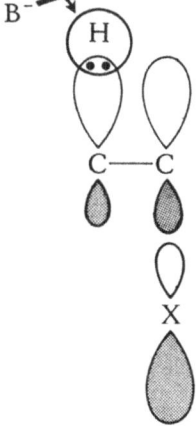

Owing to this rigid bicyclic ring system, H_1 and X cannot achieve the desired *trans*-coplanar geometry (i.e., the H_1/C/C/X angle is 120°, not 180°). Thus, one observes only *syn* elimination in which DX is ejected (the D/C/C/X angle being 0°). Apparently, the angle preference for E2 elimination is 180° > 0° > 120°. MO diagrams nicely explain this sequence:

180° (most favorable)

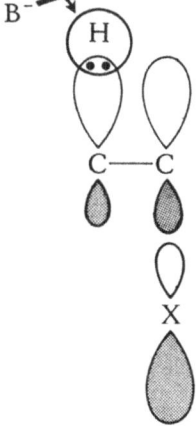

0° (possible but infrequent)

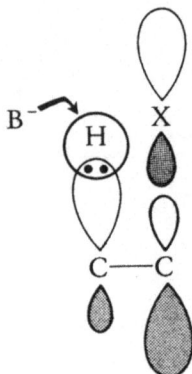

120° (no elimination)

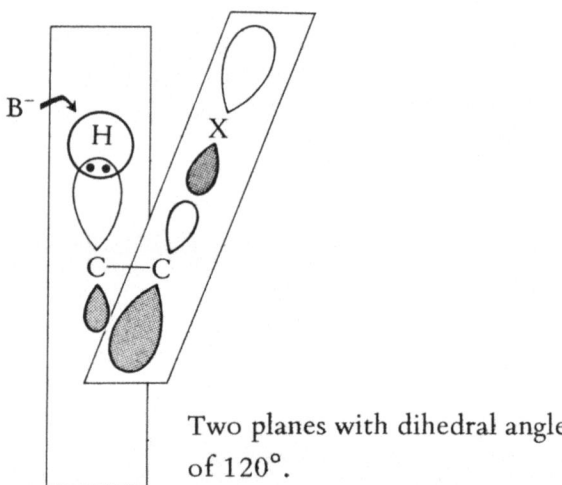

Two planes with dihedral angle of 120°.

For the sake of simplicity, the base orbital has been omitted from these pictures; the base would have a filled orbital matching that of the proton to which it ultimately bonds.

The 180° geometry is the most favorable because of perfect matching between the filled bonding C—H MO and the relevant lobes of the empty antibonding C—X MO. Note that this reaction is similar to the S_N2 back-side displacement except that in the latter the electron pair is supplied by an external species whereas in the elimination reaction the electrons derive from a C—H bond.

The MO description of a 0° geometry for elimination includes both a match and a mismatch in the transition state, thereby explaining the sluggishness accompanying a *syn* elimination. No reaction at all occurs with a 120° relationship because the orbitals cannot overlap well at this angular disposition.

Wagner–Meerwein reactions are mechanistically related to eliminations:

$$CH_3-\underset{\underset{CH_3}{|}}{\overset{\overset{CH_3}{|}}{C}}-CH_2OH \xrightarrow{H_2SO_4} \overset{CH_3}{\underset{CH_3}{}}C=C\overset{CH_3}{\underset{H}{}}$$

A methyl migrates with its pair of electrons to an sp^2-hybridized carbonium ion:

$$CH_3-\underset{\underset{CH_3}{|}}{\overset{\overset{CH_3}{|}}{C}}-CH_2\overset{\curvearrowleft}{O}H_2 \longrightarrow CH_3-\underset{\underset{CH_3}{|}}{\overset{\overset{CH_3}{|}}{C}}-\overset{+}{C}H_2 \longrightarrow$$

$$CH_2-\overset{+}{\underset{\underset{CH_3}{|}}{C}}-CH_2-CH_3$$

Proton loss from the tertiary carbonium ion gives the final product. According to MO considerations, the migration is an allowed reaction:

$$(CH_3)_2C\!-\!\overset{+}{C}H_2$$

An empty p orbital matches the filled bonding σ orbital. (We have represented the migrating methyl group here as a circle; this simplification of the carbon–carbon bonding orbital does not change any matching relationships.)

Actually, it is unlikely that the water molecules depart prior to migration because this creates an unstable primary carbonium ion. There is no problem, however, postulating simultaneous departure and migration as long as one permits a *trans* relationship:

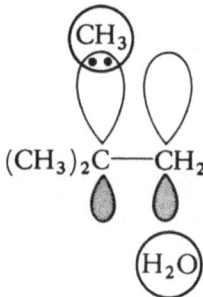

$$(CH_3)_2C\!-\!CH_2$$

The filled carbon–carbon bonding orbital serves as the donor and the empty carbon–oxygen antibonding orbital serves as the acceptor in this reaction.

The Stevens and Wittig rearrangments exemplify an entirely

different type of reaction, namely, migration to an *anionic* site:

$$\underset{\displaystyle \overset{|}{\underset{+}{\text{C}_6\text{H}_5\text{COCH}}-\overset{\displaystyle \text{CH}_2\text{C}_6\text{H}_5}{\text{N}}(\text{CH}_3)_2}}{} \longrightarrow \underset{\displaystyle \overset{|}{\text{C}_6\text{H}_5\text{COCH}-\text{N}(\text{CH}_3)_2}}{\overset{\displaystyle \text{CH}_2\text{C}_6\text{H}_5}{}}$$

$$\text{RCH}_2-\text{O}-\text{R}' \xrightarrow{\text{BuLi}} \overset{-}{\text{RCH}}-\text{O}-\text{R}' \longrightarrow \underset{\displaystyle \overset{|}{\text{R}'}}{\text{RCH}-\text{O}^-}$$

The temptation is to propose a simple one-step mechanism:

$$\underset{\displaystyle \overset{|}{\underset{+}{\text{C}_6\text{H}_5\text{COCH}}-\overset{\displaystyle \curvearrowright\text{CH}_2\text{C}_6\text{H}_5}{\text{N}}(\text{CH}_3)_2}}{} \longrightarrow \underset{\displaystyle \overset{|}{\text{C}_6\text{H}_5\text{COCH}-\text{N}(\text{CH}_3)_2}}{\overset{\displaystyle \text{CH}_2\text{C}_6\text{H}_5}{}}$$

But MO theory suggests that such a mechanism is unlikely. We can consider the reaction as an interaction between an occupied sp^3 orbital of the carbanion with an empty antibonding orbital:

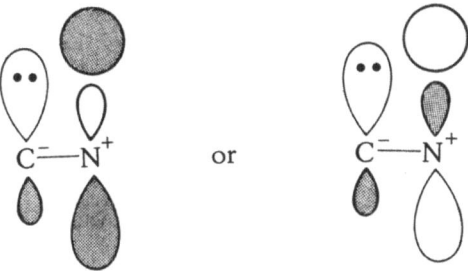

Both formulations include a mismatch at the reactive center (similar to the situation with front-side S_N2 reactions). Note that it makes no difference to our conclusion whether the carbanion has its free electron pair in an sp^3 or p orbital. A more likely mechanism for the rearrangement thus entails *two* steps such as

$$\overset{\displaystyle \overset{\frown}{C}H_2C_6H_5}{C_6H_5COC\overset{\frown}{H}-\underset{+}{N}(CH_3)_2} \longrightarrow C_6H_5COCH=\underset{+}{N}(CH_3)_2$$

$$\overset{\displaystyle CH_2C_6H_5}{\longrightarrow C_6H_5CO\overset{|}{C}H-N(CH_3)_2}$$

A similar process involving radical intermediates instead of ionic intermediates is also possible.

So far we have discussed only single bonds, but unsaturated functionalities can be treated in the same manner. Overlap of the two p orbitals in ethylene leads to a carbon–carbon double bond comprised of bonding (π) and antibonding (π^*) MO's:

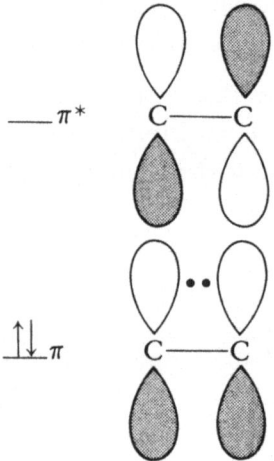

The shape of these molecular orbitals should not be taken too literally (this is true of *all* MO's in this book). For example, the bonding MO of the π bond would be better pictured as

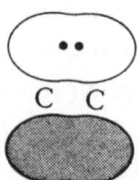

However, it is simpler to represent the MO's in terms of the component p orbitals; overlap is implied by the matching of colors.

Let us now analyze electrophilic attack on carbon–carbon double bonds as illustrated by addition of HBr to ethylene:

$$CH_2{=}CH_2 + HBr \longrightarrow CH_3{-}CH_2Br$$

Initial interaction of the filled π MO with the empty s orbital of H^+ is allowed:

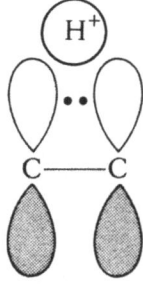

Reaction of the complex with bromide ion (gray circle) generates the product:

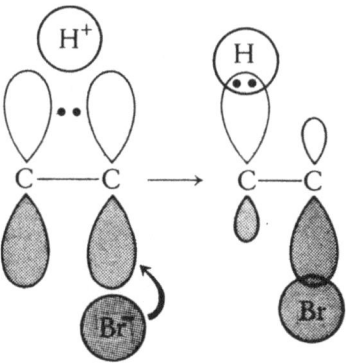

When the double bond is not substituted symmetrically, the bromine generally bonds to the more alkylated carbon (Markovnikov's rule):

$$CH_2=\overset{\overset{\displaystyle H}{|}}{C}-CH_3 \xrightarrow{\text{HBr}} CH_3-\overset{\overset{\displaystyle H}{|}}{\underset{\underset{\displaystyle Br}{|}}{C}}-CH_3$$

This behavior can be explained by assuming that the methyl group *perturbs* the MO's. Since the methyl group is slightly electron donating, the *filled* π orbital will be distorted away from the methyl group:

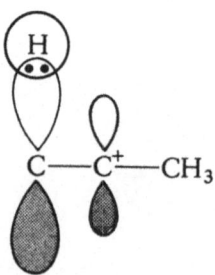

It is as if the methyl group acts like a negatively charged center, thereby repelling the electron cloud bearing two electrons away from it. Owing to this distortion, the proton prefers to overlap near the terminal carbon where the electron density is highest:

Bromine then bonds to the central carbon, giving the Markovnikov product.

Concerted *cis* addition of HBr is not facile; this is true whether one uses the double bond as a donor (π MO) or as an acceptor (π^* MO):

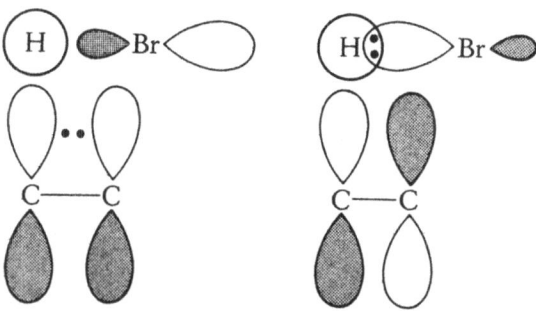

Bromination of an alkene is believed to involve a bromonium ion:

$$CH_2\!=\!CH_2 + Br_2 \longrightarrow CH_2\!\overset{\overset{\displaystyle Br}{\diagdown^+\diagup}}{\cdots}CH_2 + Br^-$$

The bromonium ion is then attacked by bromide to give the dibromide:

$$CH_2\!\overset{\overset{\displaystyle Br}{\diagdown^+\diagup}}{\cdots}CH_2 \longrightarrow \overset{\overset{\displaystyle Br}{|}}{CH_2}\!-\!\overset{\underset{\displaystyle Br}{|}}{CH_2}$$
$$\underset{\displaystyle Br}{\nwarrow}$$

The bromonium ion can be viewed as a three-center bond formed by nucleophilic displacement of bromide ion from bromine by the

π electrons of the double bond:

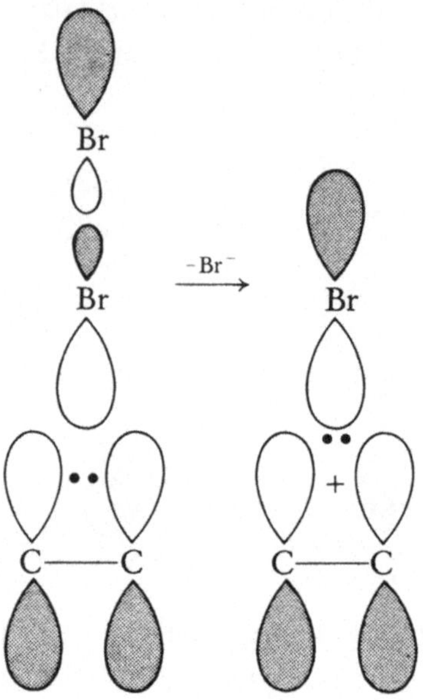

If one allows borane (BH_3) to react with 1-methylcyclo-pentene, *anti-Markovnikov* product results. Moreover, the structure of the product manifests *cis* addition:

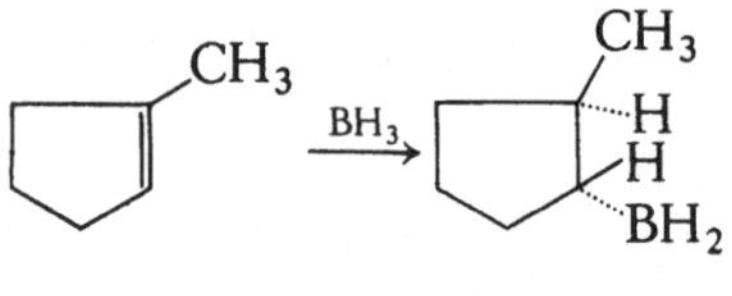

How can we explain this behavior which is totally different from addition of HBr to double bonds? The key factor here is that the boron in BH_3 has only six outer electrons and therefore possesses an empty p orbital. When borane reacts with an asymmetrically substituted double bond as in propene, the boron bonds to the carbon with the larger electron density:

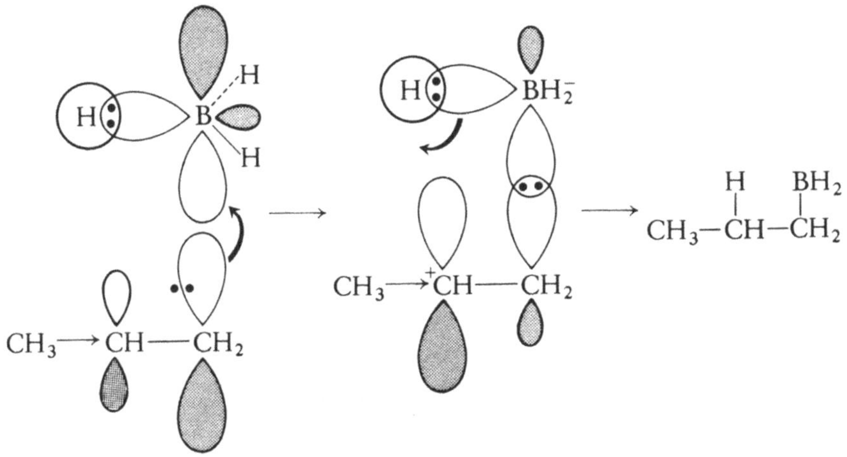

The above mechanism is shown in two steps with a carbonium ion intermediate, but it could have been written just as well in a cyclic concerted process:

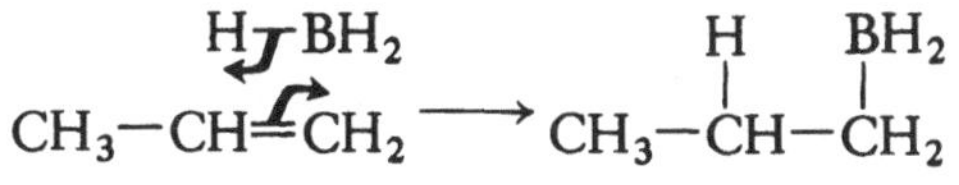

$$CH_3-CH\!\!=\!\!CH_2 \xrightarrow{} CH_3-\underset{|}{\overset{H}{C}}H-\underset{|}{\overset{BH_2}{C}}H_2$$

Either way the orbitals explain why the boron and hydrogen add *cis* to each other in contrast to addition of HBr.

Nucleophiles are capable of reacting rapidly with carbonyl groups but not with isolated carbon–carbon double bonds:

$$HO^- + H_2C\!\!=\!\!O \longrightarrow HO-CH_2-O^-$$

$$HO^- + H_2C\!\!=\!\!CH_2 \longrightarrow \text{no reaction}$$

One can view nucleophilic addition to unsaturated carbons as an interaction between an electron donor (the nucleophile) and an electron acceptor (the empty π^* orbital of the double bond). Orbital considerations predict that addition to a carbon–carbon double bond might be possible although a mismatch between the nucleophile and one of the π^* lobes would probably impede the process to a certain extent:

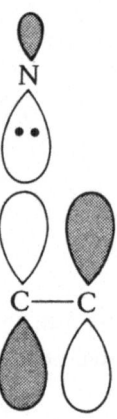

Apparently, nucleophilic attack on a carbon–carbon double bond could in principle occur were it not for two nonorbital factors: (a) An electron-rich nucleophile is electrostatically reluctant to react with an electron-rich double bond. (b) Addition to the double bond generates a high-energy carbanion unstabilized by resonance.

Neither of the two objections applies to carbonyl addition. The carbonyl has an electron-poor carbon owing to inductive withdrawal by the oxygen. Moreover, addition of a nucleophile places the negative charge on an oxygen, not a carbon. Interestingly, the MO method also predicts a favored carbonyl addition reaction. The carbonyl MO's can be represented as shown below:

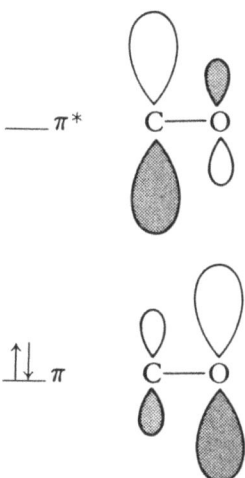

The orbitals are seen to be distorted according to what one might expect from electronegativity differences between oxygen and carbon. In the filled bonding MO, the oxygen has the larger lobe since it possesses the greater share of the π electrons. On the other hand, the antibonding MO is empty and provides no electron density between the oxygen and carbon atoms; in this situation the carbon actually has the *larger* lobe. The large lobe on carbon in the π^* MO favors overlap with a filled orbital of the nucleophile:

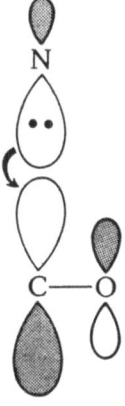

The mismatch between the nucleophile and the oxygen lobe is, in addition, diminished relative to the corresponding mismatch in the carbon–carbon double bond.

Note that the nucleophile attacks to maximize overlap and that the following mode would be forbidden:

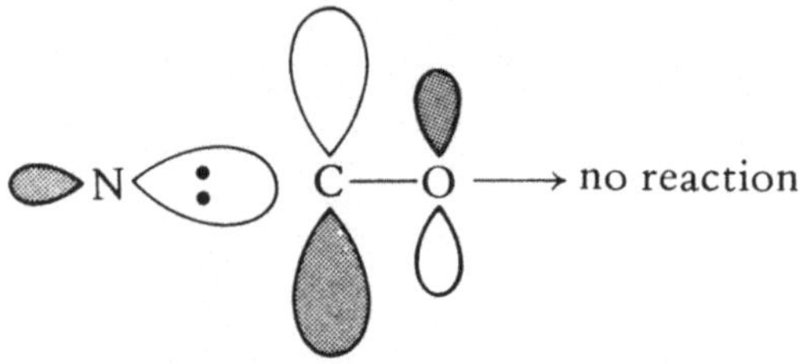

In the above representation, the nucleophile approaches the carbonyl at a node in the molecular orbital where the electron density is zero. Thus, this is a nonproductive geometry. We conclude, therefore, that nucleophilic addition to a carbonyl occurs with the nucleophile *above* the plane of the carbonyl rather than in the plane:

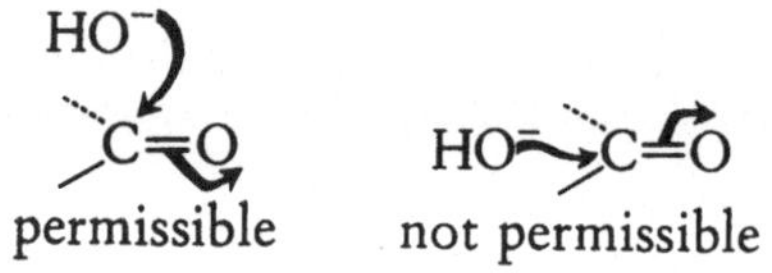

In support of this conclusion, it has been found that basic hydrolysis of the following lactone (where in-plane attack is sterically impossible) occurs rapidly:

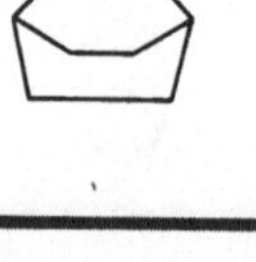

Normally, [2 + 2] cycloadditions (such as dimerization of ethylene to cyclobutane) do not occur under thermal conditions:

$$\begin{matrix} CH_2 & CH_2 \\ \| & \| \\ CH_2 & CH_2 \end{matrix} \xrightarrow[\text{heat}]{\times} \begin{matrix} CH_2-CH_2 \\ | \quad\quad | \\ CH_2-CH_2 \end{matrix}$$

("Electron pushing" does not explain this fact, illustrating again how necessary it is to use *both* electron pushing and MO theory to interpret organic reactions.) The inertness of alkenes toward thermal cycloadditions can be understood in terms of the reacting MO's. A filled bonding orbital does not match an empty antibonding orbital:

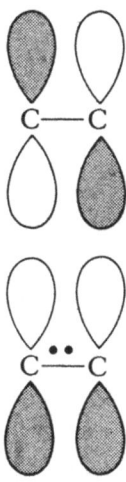

Thus, a *concerted* cycloaddition is not possible. Of course, one cannot exclude a *two-step* mechanism, but such a process, although allowed, involves an energetic diradical intermediate:

$$\begin{matrix} CH_2 & CH_2 \\ \| & \| \\ CH_2 & CH_2 \end{matrix} \longrightarrow \begin{matrix} CH_2-CH_2 \\ | \quad\quad | \\ CH_2\cdot \quad \cdot CH_2 \end{matrix} \longrightarrow \begin{matrix} CH_2-CH_2 \\ | \quad\quad | \\ CH_2-CH_2 \end{matrix}$$

One might ask why two ethylenes could not approach each other in perpendicular planes in order to achieve proper matching for cyclization:

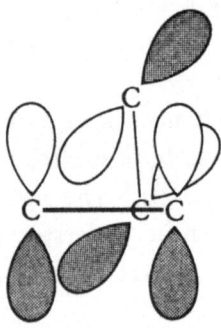

The problem here is that the orbital overlap is poor; the carbons are also not well positioned for generating the product.

Double bonds highly substituted with electron-withdrawing substituents have been observed to undergo thermal cycloadditions:

$$2\,CF_2{=}CF_2 \longrightarrow \begin{array}{c} F_2C{-}CF_2 \\ |\quad\ | \\ F_2C{-}CF_2 \end{array}$$

The explanation for this observation requires us to introduce the concept of the *correlation diagram*. A correlation diagram is a somewhat more sophisticated, but still simple, way of treating MO's. One begins by determining the symmetries of the orbitals in the reactants and in the products. If the occupied orbitals of the reactants are "smoothly correlated" with the occupied orbitals of the product, then the reaction will be allowed. It will become clear shortly, using a specific example, what "smoothly correlated" means. We should point out here, however, that a smooth correlation of symmetry does not automatically mean that the reaction will be fast; it just means that the reaction is allowed and will be faster than a similar process which is *not* correlated. Many factors, including bond strengths, electrostatic effects, steric effects, entropy effects, etc., can adversely affect a reaction even if MO theory permits the reaction.

In the derivation of a correlation diagram, symmetry properties of the p orbitals of the two alkenes are first assigned with respect to two perpendicular planes (P_1 and P_2):

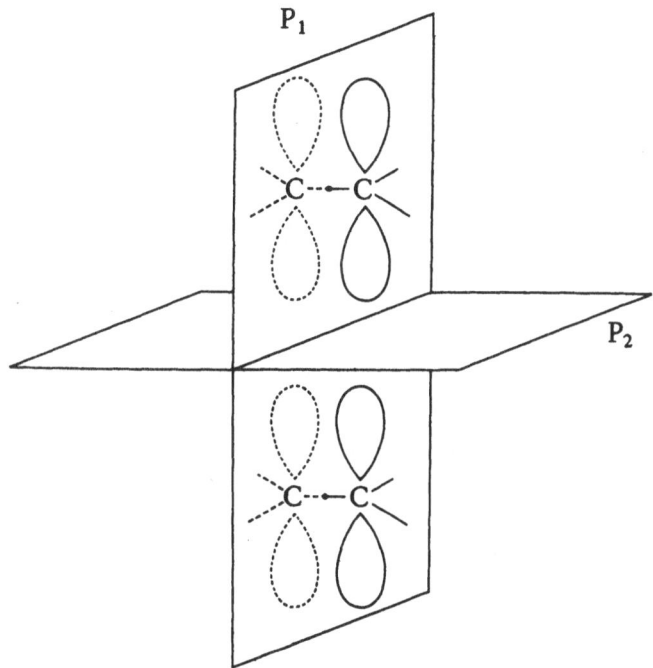

P_2 is parallel to the planes of the alkenes whereas P_1 is perpendicular to the carbon–carbon bonds and slices the alkenes into two equal portions. We will use an "S" to indicate that a set of orbitals is symmetric about P_1 or P_2 and an "A" to indicate that the set is antisymmetric about P_1 or P_2. For the two alkenes there are four orbital combinations with different symmetries:

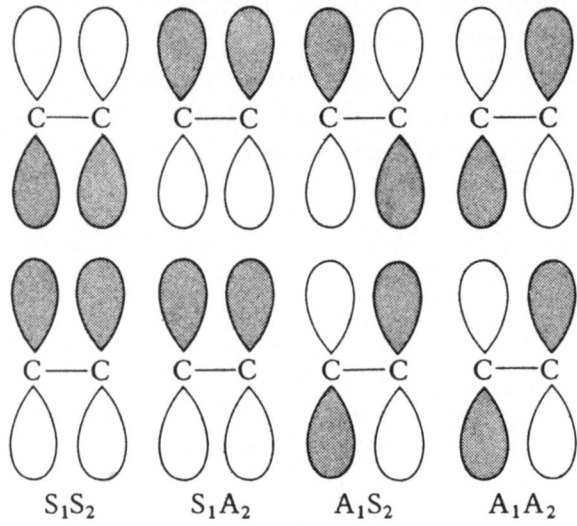

$$S_1S_2 \qquad S_1A_2 \qquad A_1S_2 \qquad A_1A_2$$

(There are, of course, other permutations, but we have restricted the "mixing" of the two orbital systems to orbitals of similar energy, i.e., bonding with bonding and antibonding with antibonding.) It is seen that the first combination is symmetric about both P_1 and P_2 and hence is assigned a symmetry of S_1S_2. The third combination is antisymmetric about P_1 but symmetric about P_2 and hence is A_1S_2. To tell whether a combination is symmetric or not about a plane, pretend that the plane is a mirror. If the orbitals on one side of the mirror "see" their reflection exactly, the combination is symmetric about that plane. The relative energies of the four combinations are easy to assess by counting the number of matches. Clearly, S_1S_2 (which gives a total of six matches about P_1 and P_2) is a lower energy combination than A_1A_2 (which lacks any matches). The four energy levels, two of which are filled with the four π electrons, are shown below:

$$\underline{\quad\quad}\quad A_1A_2$$

$$\underline{\quad\quad}\quad A_1S_2$$

$$\underline{\uparrow\downarrow}\quad S_1A_2$$

$$\underline{\uparrow\downarrow}\quad S_1S_2$$

The cycloaddition product, a cyclobutane, is analyzed in a similar manner focusing on the bonds that are formed in the course of the dimerization:

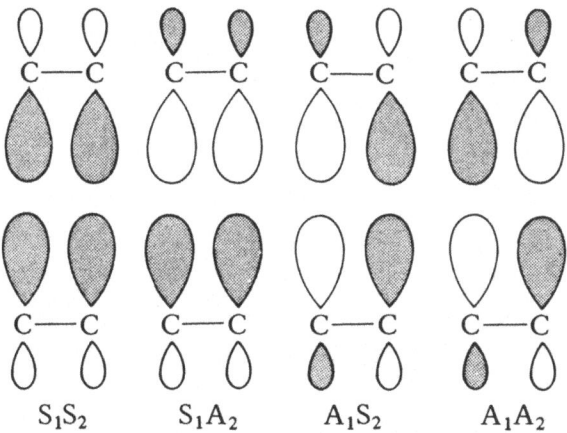

As is seen from the matching relationships in these four diagrams, S_1S_2 and A_1S_2 represent bonding σ orbitals whereas S_1A_2 and A_1A_2 constitute antibonding situations. The relative energies of the four levels can be assessed by counting the number of matches among the larger lobes (those leading to cyclobutane).

We are now in a position to correlate the symmetry of the reactant orbitals with that of the product orbitals:

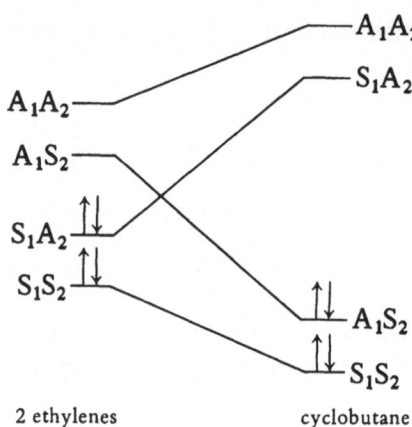

2 ethylenes   cyclobutane

Note first of all that the reactant and product energy levels come in two sets: bonding (filled) and antibonding (empty). It is reasonable that there should be a smaller energy difference between two bonding levels than between a bonding and an antibonding level. Furthermore, the bonding electrons of cyclobutane are shown to be lower in energy than the bonding electrons in the ethylenes (consistent with the fact that a σ bond is more stable and less reactive than the corresponding π bond).

The correlation diagram has tie lines which join reactant and product levels of identical symmetry. These tie lines show that the reactant orbitals do not go over "smoothly" to product orbitals. For example, the *filled* S_1A_2 of the ethylenes corresponds to an *empty* S_1A_2 of cyclobutane. The *empty* A_1S_2 of the ethylenes corresponds to the *filled* A_1S_2 of the product. In a smooth correlation, a reactant with a filled S_1A_2, for example, would always be likewise filled in the product. The cycloaddition reaction does indeed correlate with S_1S_2, but for a facile reaction *all* filled levels must do so. We conclude that thermal [2 + 2] cycloadditions are disfavored.

One could have achieved a smooth correlation (that is, "conserved symmetry") by placing two σ electrons of the product in

S_1A_2:

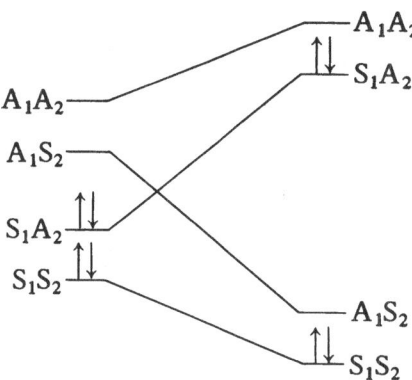

However, the energy required to elevate two electrons from a bonding orbital to an antibonding orbital would be unacceptably high.

In contrast to thermal [2 + 2] cycloadditions, *photochemical* [2 + 2] cycloadditions occur readily. A correlation diagram shows why this is so; light induces an excitation of a π electron from an S_1A_2 level to an A_1S_2 level to thus provide a smooth correlation from reactant to product:

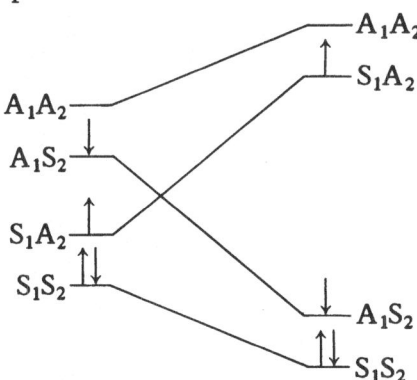

Symmetry is conserved, and the reaction is allowed. Note, however, that the product is formed as an excited state; the product quickly relaxes to the ground state with the release of thermal energy.

We now return to the observation that initiated this entire discussion of correlation diagrams, namely, that tetrafluoroethylene *does* undergo a [2 + 2] cycloaddition. It turns out that the fluorines on the double bond have the effect of raising the S_1A_2 level and lowering the A_1S_2 level. Thus, a thermal $S_1A_2 \to A_1S_2$ electron transfer becomes possible where normally only a photon could supply enough energy to induce the transition. Cycloaddition via a diradical intermediate is then permissible:

$$
\begin{array}{c}
CF_2{=}CF_2 \\
CF_2{=}CF_2
\end{array}
\xrightarrow{\text{heat}}
\begin{array}{c}
CF_2{-}CF_2 \\
\,\cdot CF_2 \quad CF_2\cdot
\end{array}
\longrightarrow
\begin{array}{c}
CF_2{-}CF_2 \\
CF_2{-}CF_2
\end{array}
$$

1,3-Dipolar additions to carbon–carbon double bonds provide one of the most important means of synthesizing heterocycles. The 1,3-dipolar compound can be either of two types:

(1)
$$\overset{+}{A}{=}\underset{\displaystyle \cdot\cdot}{B}{-}\bar{C} \longleftrightarrow A{\equiv}\overset{+}{B}{-}\bar{C}$$

$$
\begin{array}{c}
\overset{+}{N}\!\!\diagup\!\!\overset{\displaystyle N}{\diagdown}\!\!\bar{C}R_2 \\
R_2C{=}CR_2
\end{array}
\longrightarrow
\begin{array}{c}
N\!\!\diagup\!\!\overset{\displaystyle N}{\diagdown}\!\!CR_2 \\
R_2C{-}CR
\end{array}
$$

(2)
$$\overset{+}{A}{-}\underset{\displaystyle \cdot\cdot}{B}{-}\bar{C} \longleftrightarrow A{=}\overset{+}{B}{-}\bar{C}$$

$$
\begin{array}{c}
\overset{\displaystyle R}{\overset{|}{N}} \\
R_2\overset{+}{C}\!\!\diagup\quad\diagdown\!\!\bar{O} \\
R_2C{=}CR_2
\end{array}
\longrightarrow
\begin{array}{c}
\overset{\displaystyle R}{\overset{|}{N}} \\
R_2C\!\!\diagup\quad\diagdown\!\!O \\
R_2C{-}CR_2
\end{array}
$$

Other 1,3-dipolar compounds include ozone, azides, nitrile oxides, and diazoalkanes. The cycloadditions are believed to occur concertedly in agreement with the following MO analysis. We will, first of all, treat the 1,3-dipolar compound as a simple allylic system:

$$A=B-\bar{C} \longleftrightarrow \bar{A}-B=C$$

The four π electrons fill two of three MO's:

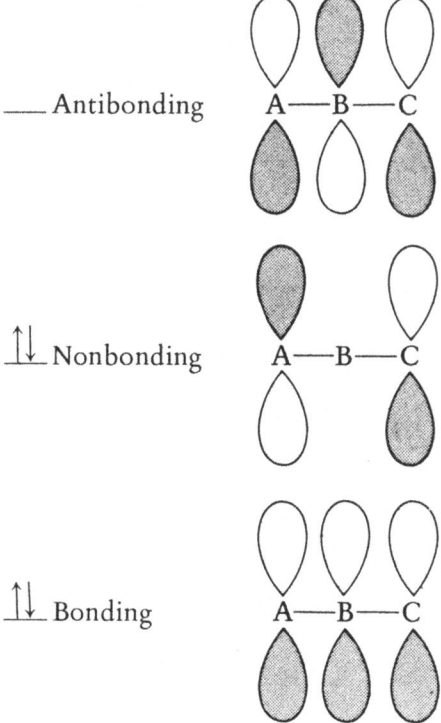

Although MO calculations are needed to prove that the three MO's are suitably represented as shown above, we can at least demonstrate that the diagrams are reasonable. It was mentioned previously that the fewer the number of nodes in an MO, the lower the energy. Thus the allylic system has a low-energy bonding MO without any nodes; the nonbonding MO has one node at atom B; and the antibonding MO has two nodes, one midway between A and B and the other midway between B and C. The nonbonding MO neither stabilizes nor destabilizes the holding of atoms together; calculations show that the p orbital of atom B contributes nothing to the mixing of p orbitals to generate the nonbonding MO.

All examples discussed thus far in this chapter have utilized an interaction between a filled MO as an electron donor and an empty MO as an electron acceptor. We now face a case where there are *two* filled orbitals. Which one do we use to interact with the empty antibonding MO of the alkene? According to "frontier orbital theory," it is the highest occupied MO (HOMO) and the lowest unoccupied MO (LUMO) orbitals that are crucial to the reaction. This is reasonable. The highest filled orbital has the most reactive and energetic electrons; the lowest empty orbital is the one which would be filled first if additional electrons became available. Thus, the 1,3-dipolar cycloaddition can be pictured as an interaction between the filled nonbonding MO of the 1,3-dipolar compound with the empty antibonding MO of the alkene:

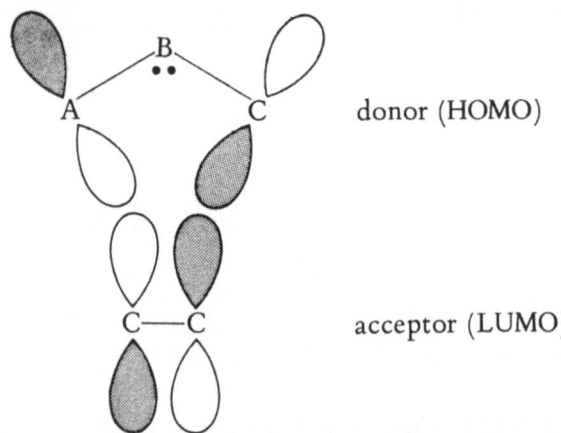

donor (HOMO)

acceptor (LUMO)

Alternatively, the reaction could involve the filled bonding MO of the alkene and the empty antibonding MO of the 1,3-dipolar compound:

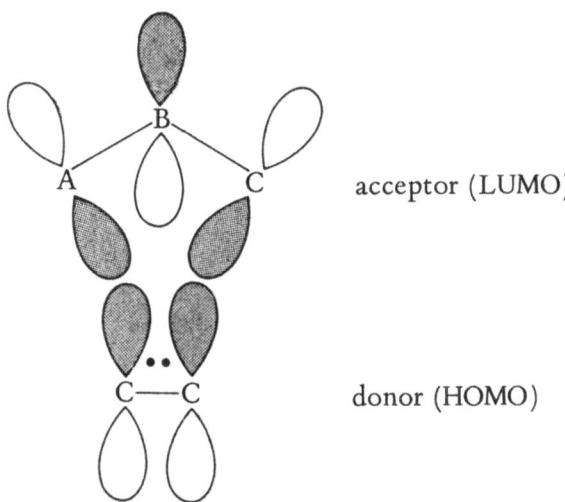

acceptor (LUMO)

donor (HOMO)

In either event the matching is satisfactory so that cycloaddition is allowed. It is important to stress again that just because a reaction is *allowed* does not mean that the reaction necessarily *will* occur. For example, 1,3-dipolar cycloaddition of an amide to an alkene is permitted by MO theory:

$$R-C \overset{O}{\underset{NH_2}{\big\backslash}} \longleftrightarrow R-C \overset{\bar{O}}{\underset{\overset{NH_2}{+}}{\big\backslash}} \quad \overset{CR_2}{\underset{CR_2}{\|}} \xrightarrow{\quad\times\quad} R-\bar{C} \overset{O-CR_2}{\underset{\overset{NH_2}{+}}{\big\backslash_{CR_2}}}$$

Yet the reaction does not occur because, among other reasons, the creation of an ylide from neutral species is energetically unfavorable. The reaction also entails loss of resonance stabilization associated with the amide.

In some cases it is useful to specify which compound acts as the donor and which acts as the acceptor (although as we saw in the 1,3-dipolar cycloaddition, this distinction does not affect the main conclusion that the reaction is allowed). It turns out that the closer two orbitals are in terms of energy, the greater the stabilization when they interact. For example, consider the following reaction:

The carbanion has a filled nonbonding MO whereas the carbonium ion is empty at this level:

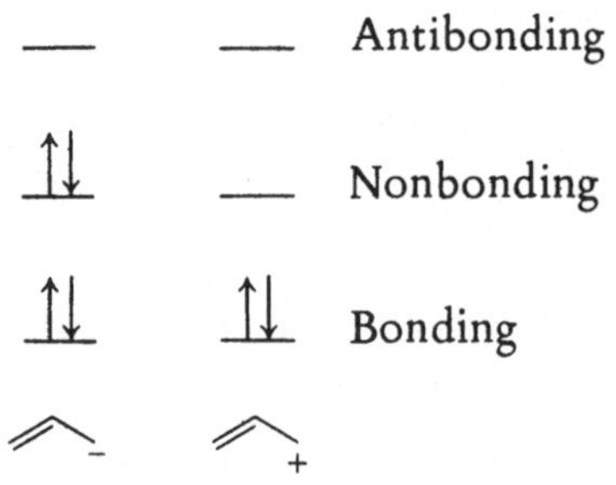

In principle one could combine the lowest unoccupied MO of the carbanion (antibonding) with the highest occupied MO of the carbonium ion (bonding). But these two molecular orbitals differ widely in energy. A more likely interaction would take place between the highest occupied MO of the carbanion and the lowest unoccupied MO of the carbonium ion (i.e., the two nonbonding orbitals). In other words, the carbanion quite reasonably acts as the nucleophile and the carbonium ion as the electrophile (and not the reverse).

The Diels–Alder reaction, used widely in synthetic organic chemistry, is an example of a [4 + 2] cycloaddition:

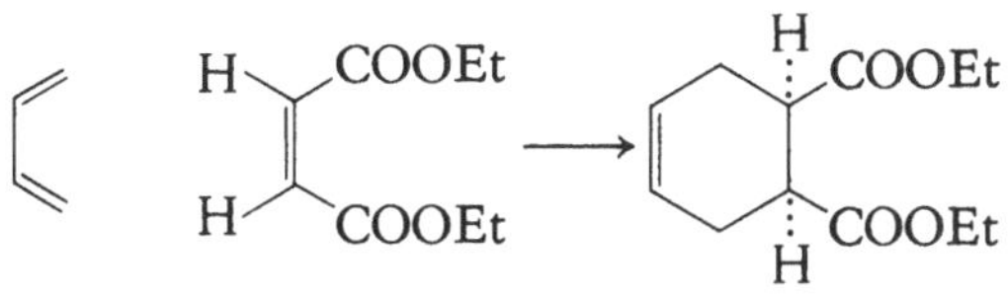

A diene reacts with a dieneophile (often a double bond conjugated with a carbonyl or some other electron-withdrawing functionality). The resulting adduct is formed stereospecifically; that is, a *cis* dieneophile gives a *cis*-disubstituted cyclohexene. For this reason and others, Diels–Alder reactions are believed to be concerted. If the two carbon–carbon bonds were formed one at a time in a non-concerted process, then a *cis* dieneophile could conceivably pro-uce *trans* product.

The four relevant MO's of a conjugated diene are shown below:

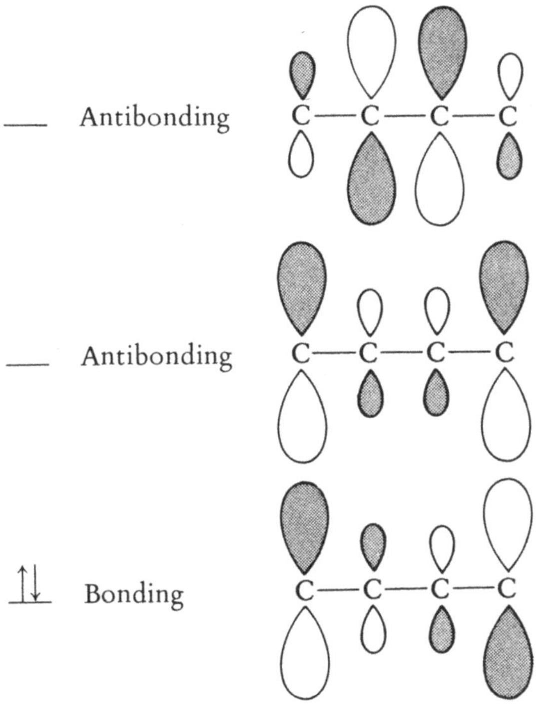

____ Antibonding   C — C — C — C

____ Antibonding   C — C — C — C

↑↓ Bonding   C — C — C — C

⇅   Bonding

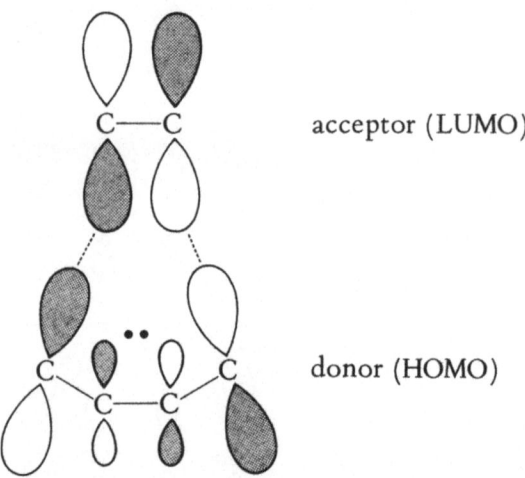

Note that the number of nodes in the MO's increases from zero in the lowest bonding MO to three in the highest antibonding MO. The size of the components of the MO are seen to be unequal; this reflects unequal contributions of the p orbitals to the MO's. For example, the second and third carbons of butadiene contribute more to the lowest bonding MO than the terminal carbons. This is not intuitive but rather comes from MO calculations which are beyond the scope of this book. The important point for our purposes, however, is that the highest bonding MO of the diene matches the empty antibonding MO of the alkene (for the sake of simplicity, the dieneophile is treated as an isolated double bond):

acceptor (LUMO)

donor (HOMO)

Reaction between a diene and a dieneophile, substituted as shown below, gives mainly the "*ortho*" product:

MO theory can explain this regioselectivity. One can show that the methoxy group perturbs the highest occupied MO of the substituted diene such that the C_4 has a larger lobe than the C_1 :

$$CH_3O-C_1-C_2-C_3-C_4$$

We ignore the lobes on C_2 and C_3 because they are not directly involved in the bond formation. Similarly, it can be shown that the unoccupied MO of the alkene has a larger lobe at C_2:

$$C_2—C_1—CHO$$

If one assumes that regioselectivity is dictated by a "large–large" interaction, then the MO diagram gives the correct product:

The "*meta*" product would be formed by two "large–small" interactions, which apparently do not compete well with a "large–large" and "small–small" overlap. Thus, MO theory can be used to rationalize the direction of a reaction, but the student should realize he does not yet have the background to make similar predictions.

"Electron pushing" predicts the same product as MO theory:

$$CH_3\overset{\curvearrowright}{O} \text{-} CH \text{=} CH \text{-} CH \text{=} CH_2 \longleftrightarrow CH_3\overset{+}{O} \text{=} CH-CH=CH-\overset{-}{C}H_2$$

$$CH_2 \text{=} CH \text{-} CH \text{=} O \longleftrightarrow \overset{+}{C}H_2-CH=CH-\overset{-}{O}$$

The molecules line up so that opposite partial charges are in proximity to each other (a reasonable postulate although there are exceptions to this):

Alternatively, the reaction can be written as if it were a two-step process even though this does not seem to be the case:

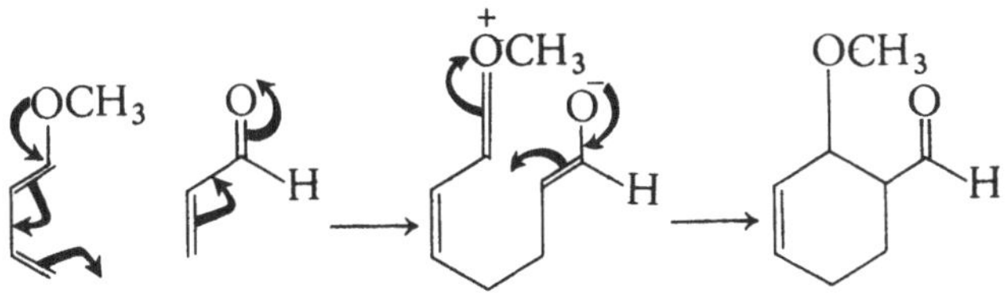

The electron pushing method shows a much higher transition state energy in the *"meta"* alignment:

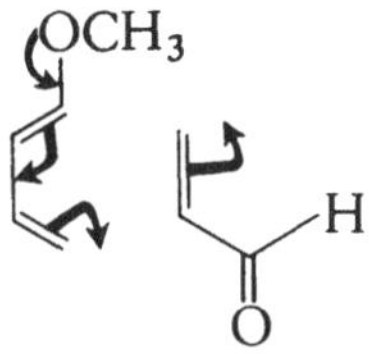

The electron pushing approach is actually more useful in predicting Diels–Alder substituent effects than the MO method because of its simplicity. In order to assess a substituent effect on regioselectivity by the MO method, one has to calculate how the substituent perturbs the orbitals; this can be a difficult task with sometimes ambiguous results.

A *pericyclic* reaction is one in which a concerted reorganization of bonding takes place throughout a cyclic array of contiguous atoms. If the pericyclic process is intramolecular, it is called an *electrocyclic* reaction. Two examples are shown below:

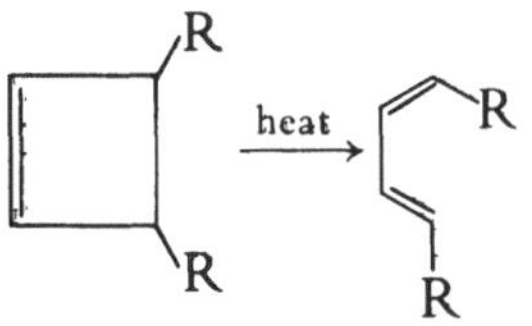

cis   *trans–cis*

trans   trans-trans

The reactions are seen to be totally stereospecific. It turns out that
the stereochemical pathway is predominantly determined by the
highest occupied MO. This is most easily illustrated by means of
the reverse reaction (diene to cyclobutene), but the ensuing concepts hold true for electrocyclic reactions in either direction. In
any event, the highest occupied MO for the diene is given below
(size differences among the lobes are ignored):

trans-cis   cis

In order to achieve proper matching in the formation of a *bonding*
σ orbital, the terminal carbons must first rotate 90° in the *same*
direction. This places the R groups of the *trans-cis* diene in a *cis*
configuration on the cyclobutene ring. Rotation in the *same* direction (both carbons clockwise or both counterclockwise) is called

conrotatory. The student should convince himself that conrotatory ring closure of the *cis–cis* diene gives the *trans*-disubstituted cyclobutene. In general, thermal electrocyclic reactions of systems with $4n$ π electrons (n being an integer) always proceed via a conrotatory mechanism.

When an electrocyclic reaction of a $4n$ system is induced by light, a *disrotatory* mode is observed; that is, the terminal carbons rotate in *opposite* directions. This is because the highest occupied orbital is now an antibonding MO owing to the excitation of an electron:

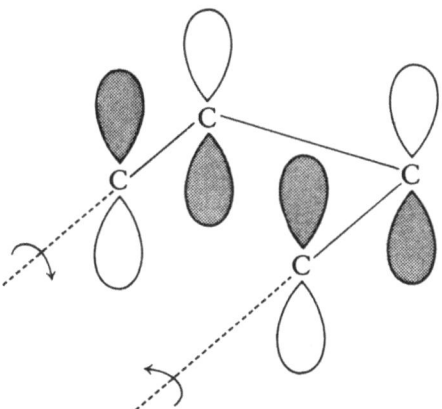

Ring closure must, consequently, occur with the terminal carbons rotating in an opposite direction to secure proper matching:

An example of a photochemical disrotatory ring closure of a $4n$ system is shown below:

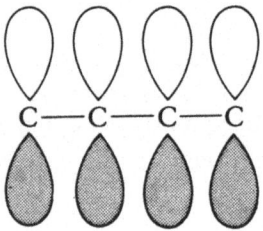

The two protons (heavy dots in product) are *cis* to each other as necessitated by the orbital considerations.

We may add parenthetically that the electrons in the double bond of the product originate from the *lowest* filled orbital. In both the thermal and photochemical electrocyclic reactions the following MO supplies the electrons needed for the new π bond:

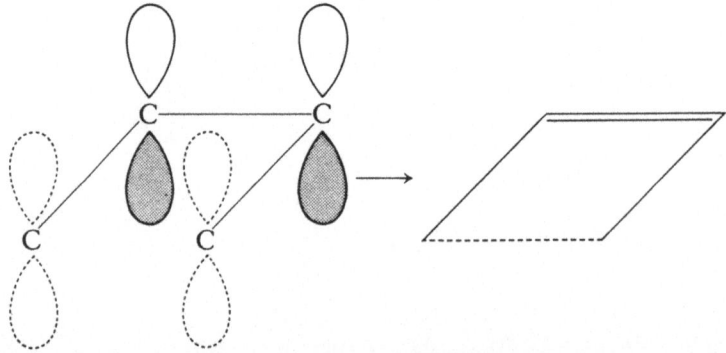

Obviously the central lobes are correctly matched for formation of the π bond:

Thermal cyclization of the following triene leads to the *cis* product:

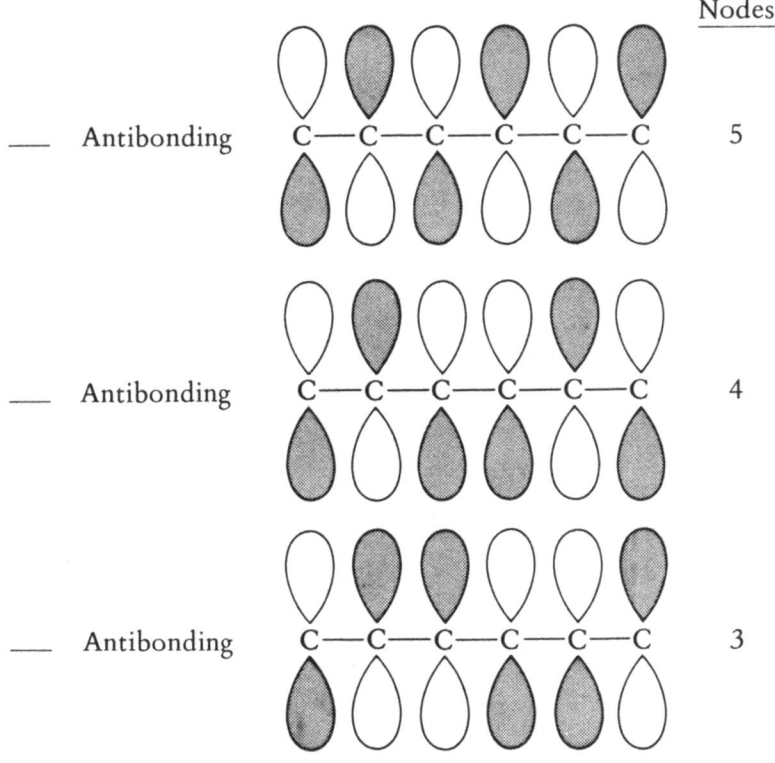

In contrast to $4n$ systems, with $[4n + 2]$ π electrons such as the above triene always engage in disrotatory thermal cyclizations. This can be understood in terms of the MO's of conjugated trienes:

		Nodes
___ Antibonding	C—C—C—C—C—C	5
___ Antibonding	C—C—C—C—C—C	4
___ Antibonding	C—C—C—C—C—C	3

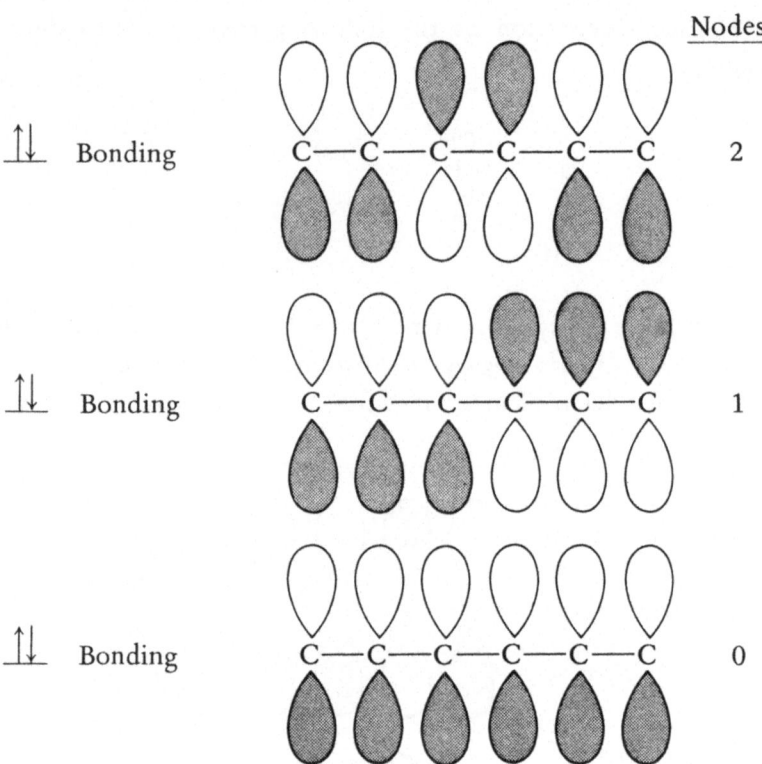

In reality the spacings between the energy levels are not identical and neither are the sizes of the lobes in the MO's; however, this has no bearing on the stereochemical conclusions. The stereochemistry is determined solely by the highest occupied MO (HOMO).

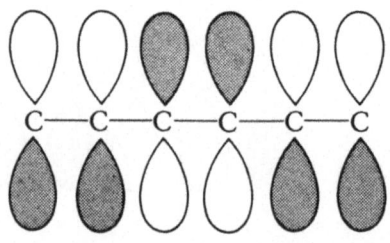

Disrotatory closure of the terminal carbons leads to proper matching:

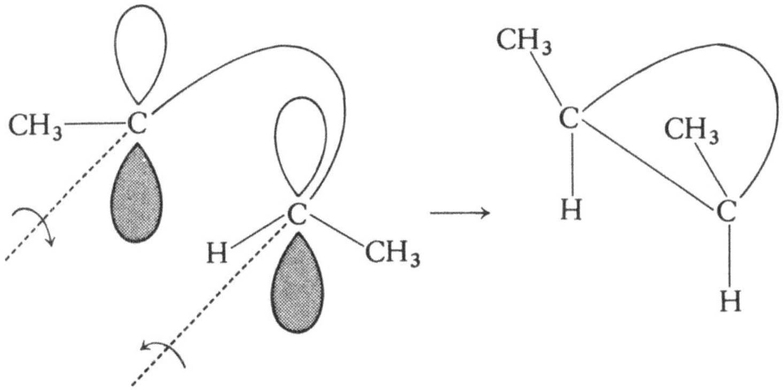

As an exercise the student should convince himself that a light-induced cyclization of a triene proceeds in a conrotatory manner. The relevant orbital in this case is, of course, the antibonding MO with three nodes.

The above concepts, embodied in the *Woodward–Hoffmann* rules, are summarized below:

Number of π electrons	Thermal	Photochemical
$4n$	Con	Dis
$4n + 2$	Dis	Con

A sigmatropic shift is a concerted rearrangement in which a singly bound atom or group moves from one terminus of a π system to the other. An example is shown below:

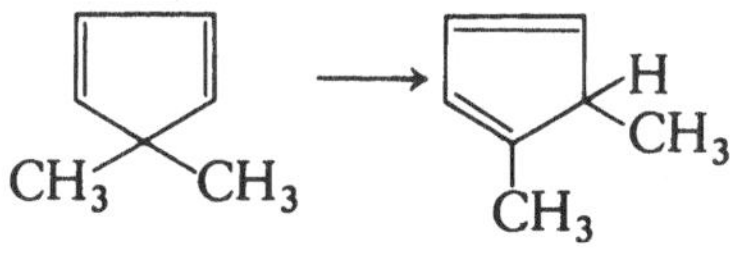

If the shift occurs on the same face of the π system, the reaction is said to be *suprafacial*; if opposite faces are involved the reaction is called *antarafacial*:

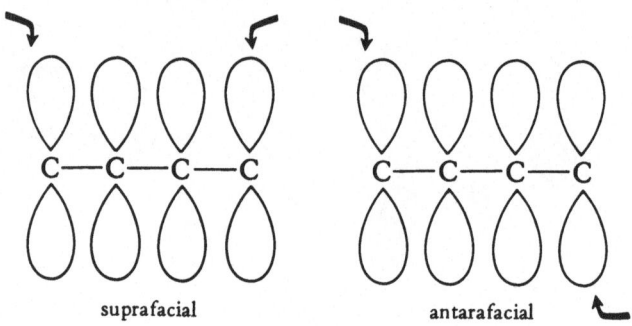

suprafacial   antarafacial

The allylic rearrangement is one of the simplest and most common types of shifts:

$$\underset{\text{C--C=C}}{\overset{\text{X}}{|}} \longrightarrow \underset{\text{C=C--C}}{\overset{\text{X}}{|}}$$

The question arises as to whether such a shift is concerted and, if so, whether it is suprafacial or antarafacial. We can formally treat the 1,3 shift as the migration of X^+ on the allyl anion:

$$X^+ \qquad\qquad\qquad X^+$$
$$\bar{C}\text{--C=C} \longrightarrow \text{C=C--}\bar{C}$$

The HOMO of the allyl anion is a nonbonding MO bearing two electrons:

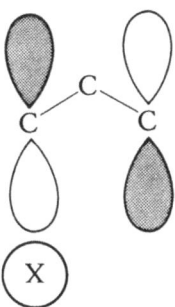

Migration of X is clearly not allowed in a suprafacial mode:

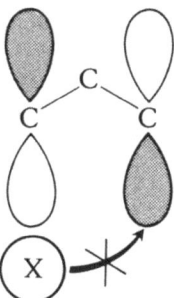

An antarafacial 1,3 shift is permissible according to MO theory:

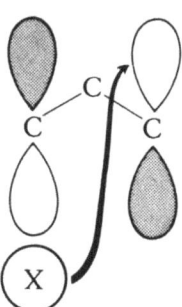

However, such a shift is sterically impossible; only when a π system is much longer can X "reach" the other face of the molecule. Thus, we conclude that a 1,3 sigmatropic shift should not occur. Why, then, are shifts of allylic halides or acetates so common? The answer is that these shifts are not concerted; instead they take place by two-step ionic mechanisms for which the rule of "highest occupied MO" does not apply:

$$\overset{\overset{\text{Cl}}{\downarrow}}{\text{C}}-\text{C}=\text{C} \xrightarrow{-\text{Cl}^-} \overset{+}{\text{C}}-\text{C}=\text{C} \longleftrightarrow \text{C}=\text{C}-\overset{+}{\text{C}} \xrightarrow{+\text{Cl}^-} \text{C}=\text{C}-\overset{\overset{\text{Cl}}{|}}{\text{C}}$$

Only concerted reactions must follow the Woodward–Hoffmann concepts.

4.2. Problems

1. Predict whether O_3, a 1,3 dipolar compound, would add to butadiene in a 1,2 or a 1,4 fashion:

$$\text{CH}_2{=}\text{CH}{-}\text{CH}{=}\text{CH}_2 + \text{O}_3 \longrightarrow \text{CH}_2{=}\text{CH}{-}\underset{\underset{\text{O}}{\overset{|}{\text{O}}}}{\text{CH}}{-}\underset{\overset{|}{\text{O}}}{\text{CH}_2}$$

or

$$\text{CH}_2{=}\text{CH}{-}\text{CH}{=}\text{CH}_2 + \text{O}_3 \longrightarrow \overset{\text{CH}=\text{CH}}{\underset{\underset{\text{O}}{\text{CH}_2 \quad \text{CH}_2}}{\diagdown \quad \diagup}}$$

2. Show that diimide can concertedly reduce an alkene. Use the alkene first as the electron acceptor and second as the electron donor.

$$R_2C{=}CR_2 + NH{=}NH \longrightarrow R_2CH{-}CHR_2 + N_2$$

3. Predictions of reaction stereochemistry from the lowest unoccupied and highest occupied orbitals of the reactants work best for exothermic reactions. Explain this statement.

4. A concerted suprafacial 1,3 sigmatropic shift is not allowed:

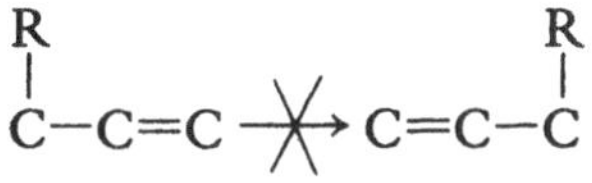

It has been pointed out, however, that keto–enol tautomerisms could in fact take place by such a shift (although they probably do not because rapid acid- or base-catalyzed tautomerisms predominate).

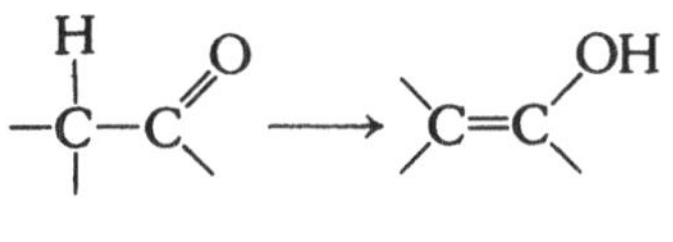

Suggest a reason why a concerted 1,3 shift might be permissible in the keto–enol equilibrium.

5. Diels–Alder addition of acrolein to cyclopentadiene could

conceivably produce *endo* or *exo* product:

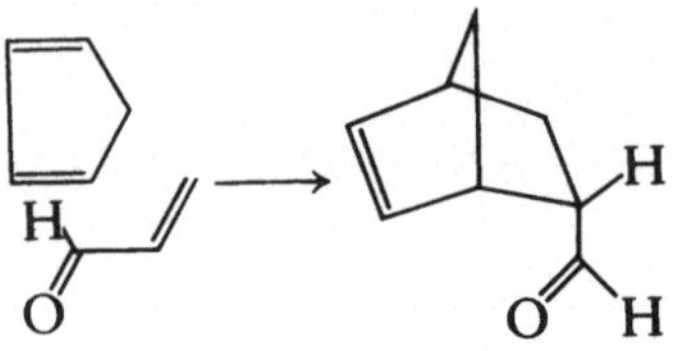

endo

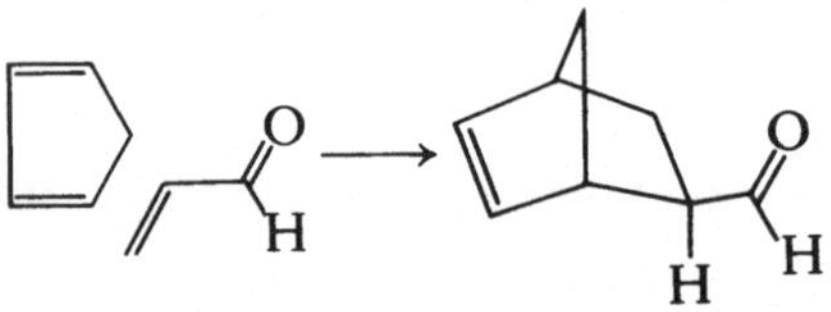

exo

Suggest a reason why the *endo* compound is in fact the major product.

---

6. Explain whether or not you would expect the cyclization of butadiene to bicyclobutane to occur concertedly.

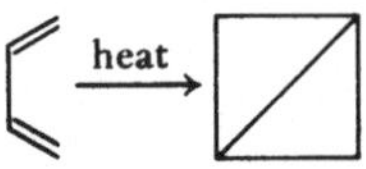

The relevant orbitals of the reactants and products are shown below:

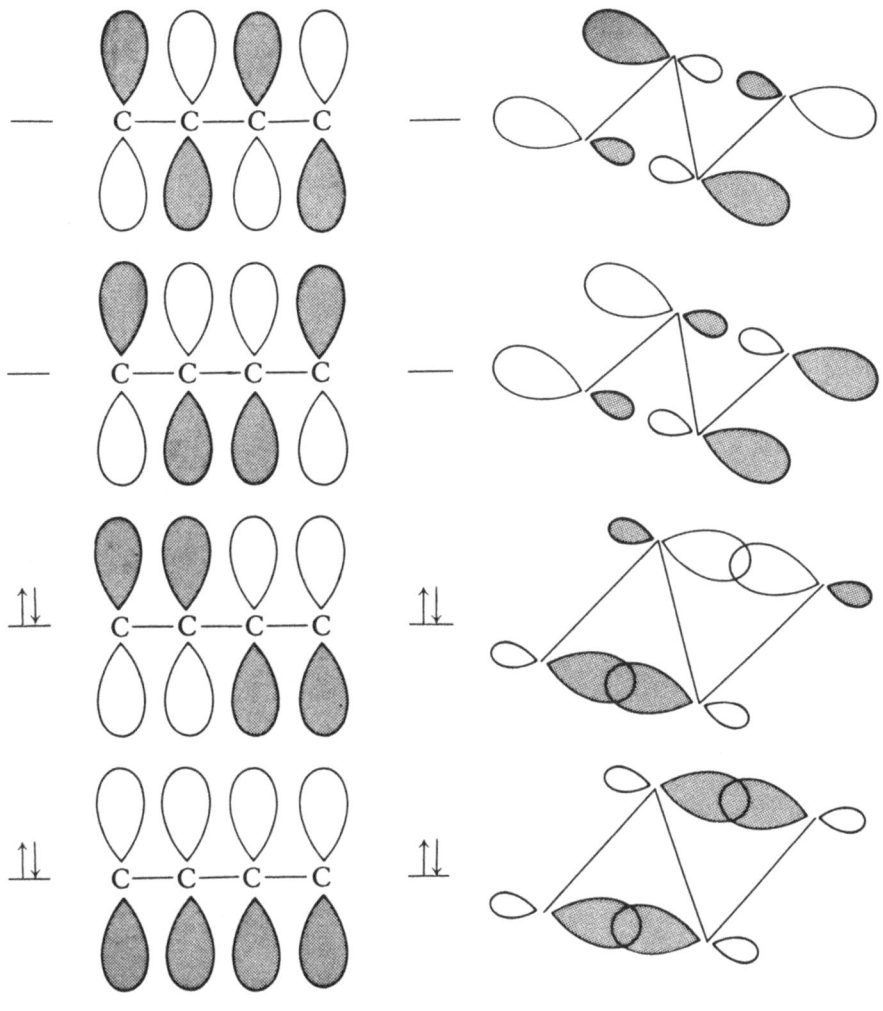

7. We have shown by means of the highest filled MO that a conjugated diene can concertedly cyclize in a conrotatory mode:

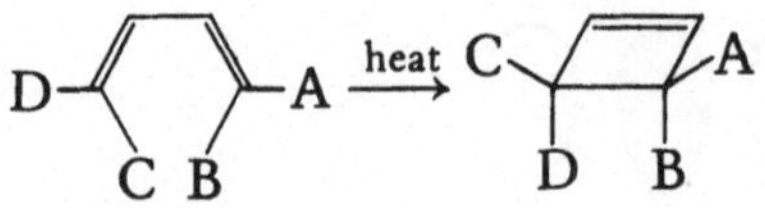

This same conclusion can be derived from a correlation diagram. The first step is to define symmetry elements. A symmetry *plane* is designated for the disrotatory mode; it is seen that a disrotatory perturbation does not affect the symmetry of the molecule with respect to the plane:

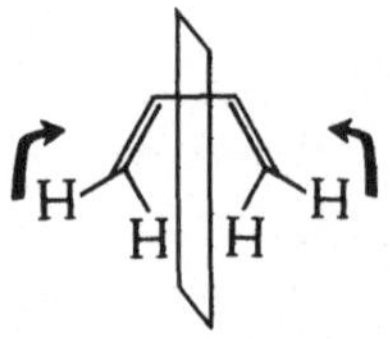

dis

A symmetry *axis* characterizes a conrotatory motion because a conrotatory perturbation does not affect the symmetry of the molecule with respect to a 180° rotation about the axis:

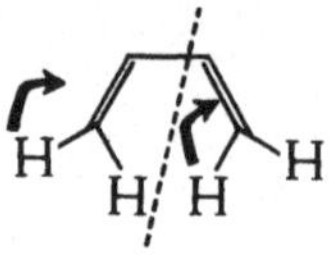

con

Thus, the following orbital can be assigned an "A" (antisymmetric) with respect to the plane but "S" (symmetric) with respect to a 180° rotation about the axis:

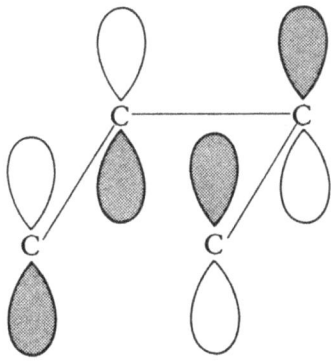

Given below are the relevant orbitals for the reactant and product. Assign first the symmetries of the orbitals with respect to the *plane*; then deduce whether or not there is a smooth correlation for a *disrotatory* closure. Next, assign the symmetries with respect to the *axis*; then deduce whether or not a *conrotatory* closure is permitted.

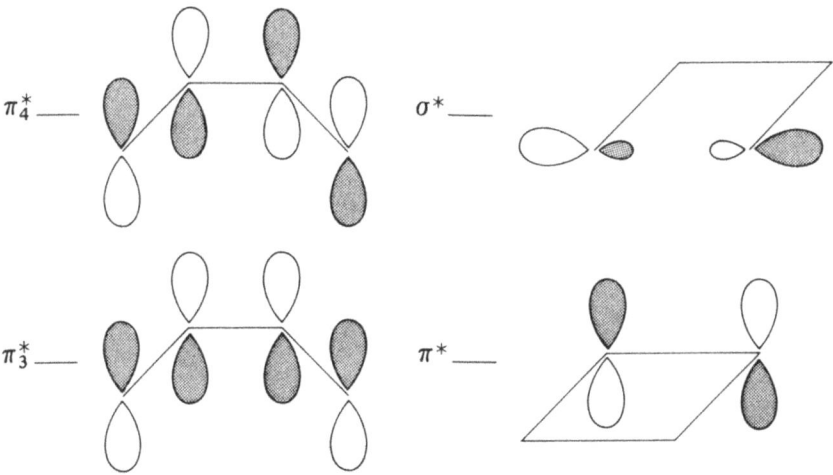

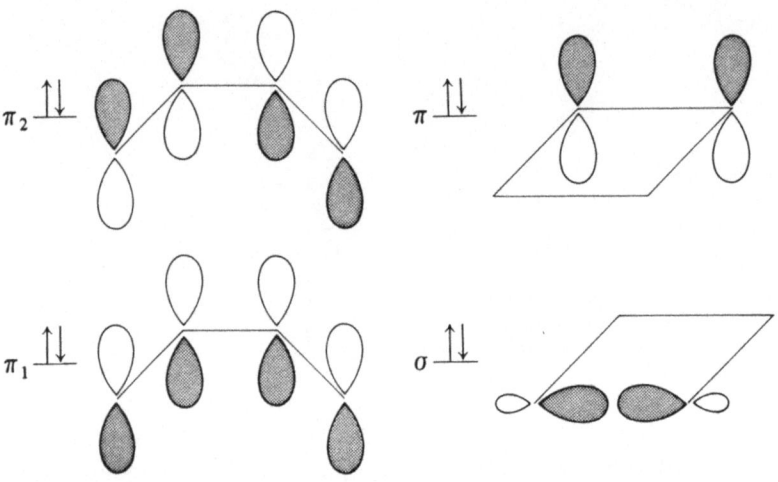

---

8. An ester enolate prefers to react with an α,β-unsaturated ester at the β carbon rather than at the carbonyl:

$$R-\underset{\underset{O}{\|}}{C}-CH_2^- \longleftrightarrow R-\underset{\underset{O}{|}}{C}=CH_2 \rightarrow CH_2=CH-\underset{\underset{O}{\|}}{C}-R$$

$$\longrightarrow R-\underset{\underset{O}{\|}}{C}-CH_2-CH_2-CH_2-\underset{\underset{O}{\|}}{C}-R$$

Suggest a reason why it might be inaccurate to rationalize this behavior using MO theory.

---

9. Nucleophilic attack on the carbonyl of unhindered cyclohexanones occurs primarily from the axial side:

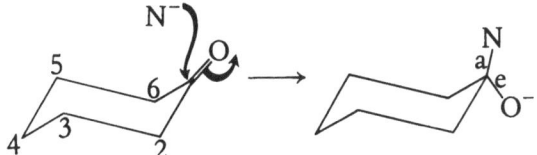

One rationalization of this behavior is that the p orbitals of the carbonyl group interact with the σ orbitals of the C_2-C_3 and C_5-C_6 carbon–carbon single bonds. Explain in more detail how such an interaction could possibly affect the side at which the nucleophile attacks.

---

10. The following is an example of an S_N2' substitution:

$$N^- \quad X \quad \longrightarrow \quad N$$
$$C{=}C{-}C \qquad C{-}C{=}C + X^-$$

Usually the nucleophile enters on the *same* side that X departs. Explain this stereochemistry assuming that it is governed by the lowest unoccupied MO of the system which is distorted such that bonding is maximized and antibonding is minimized.

---

11. Thermal electrocyclic cleavage of the following cyclobutene gives *two* products, one of which predominates in yield greatly over the other. Draw the two products and explain why there

is a difference in yield.

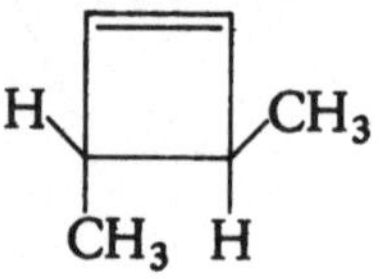

---

12. Predict the product of photochemical ring closure in the following compound:

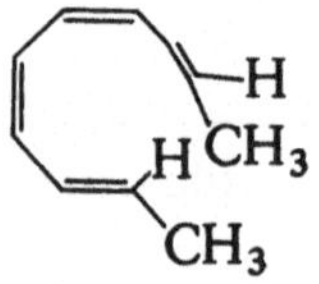

---

13. The highest occupied MO of naphthalene is shown below (top view), with the size of the circles representing roughly the electron density:

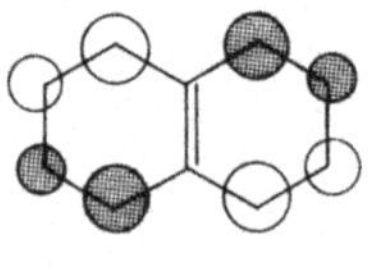

The lowest unoccupied MO of nitronium ion (NO_2^+) is also shown (side view):

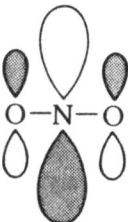

What product would you expect in the nitration of naphthalene by nitronium ion?

---

14. Explain the fact that thermal elimination of H_2 to form furan
is facile in one case but difficult in another:

$$\text{(ring)} \longrightarrow \text{(ring)} + H_2 \qquad \text{(facile)}$$

$$\text{(ring)} \longrightarrow \text{(ring)} + H_2 \qquad \text{(difficult)}$$

---

15. Predict whether or not the following reaction is allowed:

$$R_3P + XY \longrightarrow R_3P\begin{smallmatrix}X\\|\\|\\Y\end{smallmatrix}$$

Use the p orbital holding the lone pair of electrons on the
phosphorus as the highest occupied MO.

16. The highest occupied and lowest unoccupied MO's of ketene $(CH_2=C=O)$ are shown below:

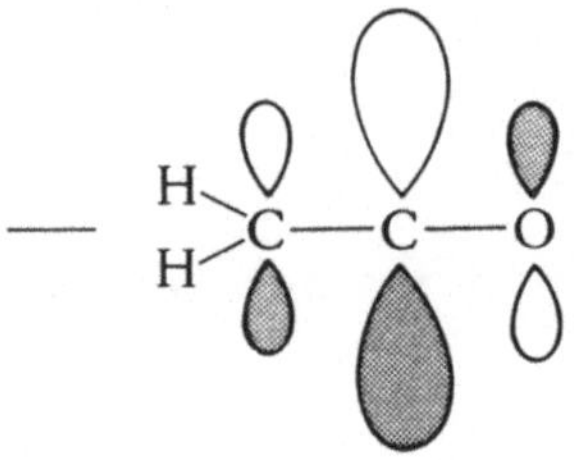

LUMO
(MO in plane of molecule)

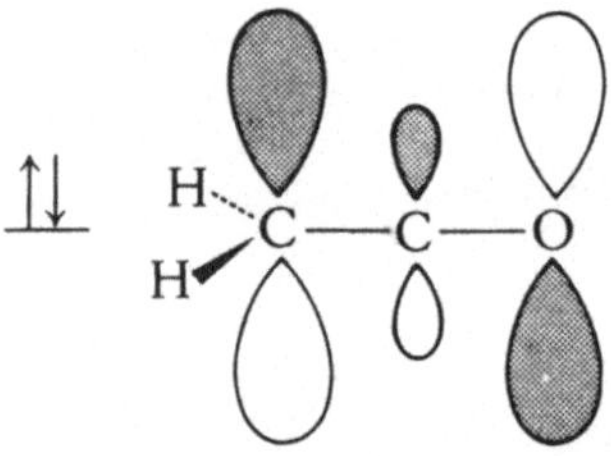

HOMO
(MO perpendicular to plane)

The two orbitals are perpendicular to each other as indicated. Predict whether the cycloaddition of ketene to carbonyl compounds gives cyclic ester or cyclic ether:

$$R_2C=O + H_2C=C=O \longrightarrow$$

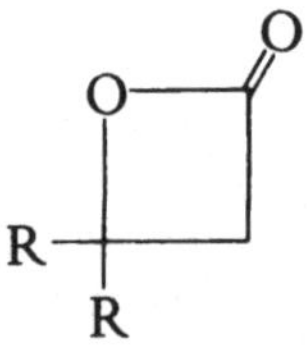

or

$$R_2C=O + H_2C=C=O \longrightarrow$$

17. The relative rates for two 1,5 sigmatropic shifts in isomeric spirononatrienes are given below:

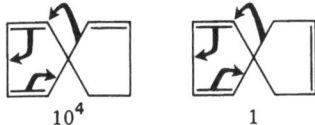

Explain the difference in terms of MO's. The two degenerate (identical energy) occupied MO's of the cyclopentadienyl system are shown below (top view):

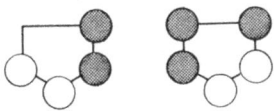

4.3. Answers

1. It would add in a 1,2 fashion:

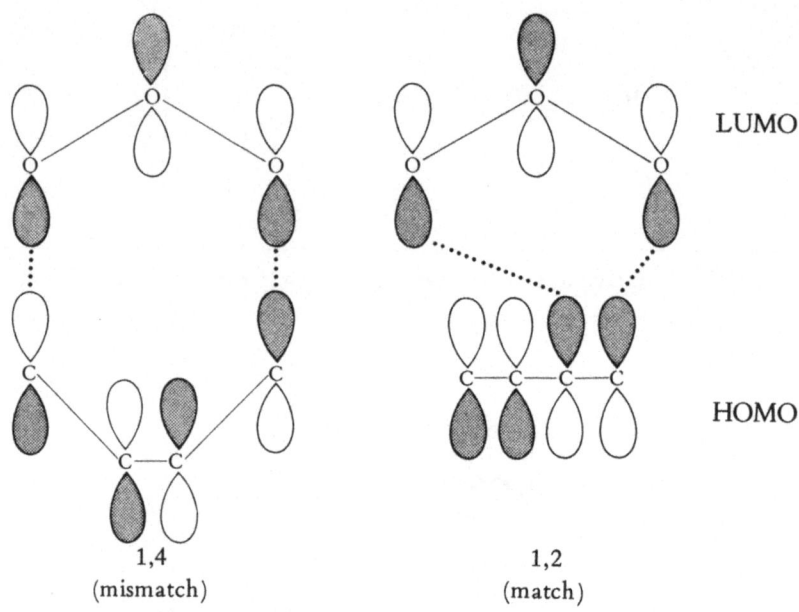

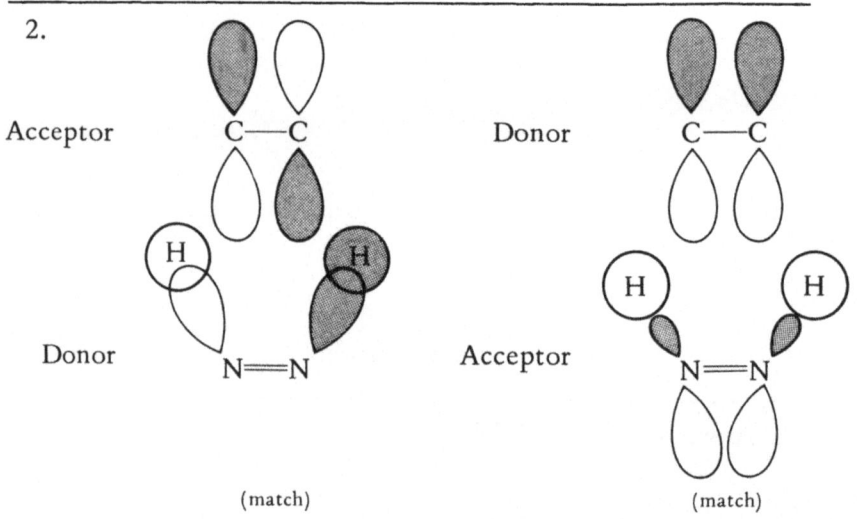

3. According to the Hammond principle, the transition state resembles the reactants for exothermic reactions and the products for endothermic reactions. Thus, orbital properties of the reactants are a better guide to the behavior of the transition state if the reaction is exothermic.

4. The oxygen has sp^2 orbitals holding unshared pairs of electrons:

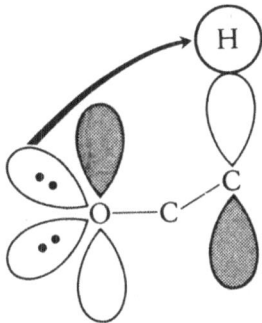

A matching can therefore be achieved.

5. In our discussion of the Diels–Alder reaction we treated the dieneophile as an isolated double bond. But to explain the preference for *endo* addition it is necessary to consider the entire π system including the carbonyl. Matching is seen to be possible with both *endo* and *exo*, but only the *endo* has an additional "secondary" matching. Hence *endo* addition predominates. In the diagrams below the cyclopentadiene is the electron donor (HOMO) and acrolein the electron ac-

ceptor (LUMO):

(1)

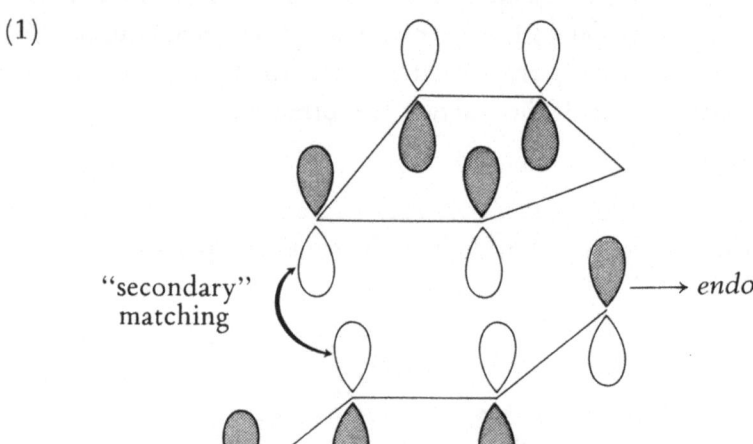

(2)

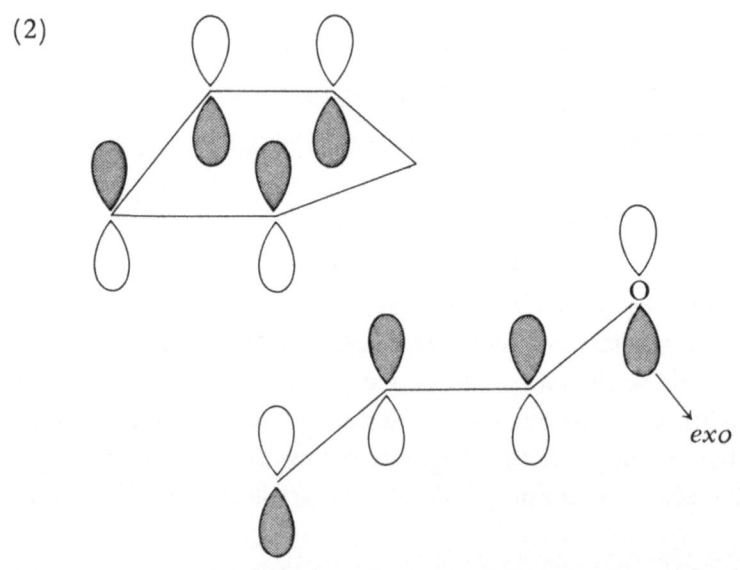

6. A concerted cyclization of butadiene to bicyclobutane would not be allowed under thermal conditions. The highest occupied MO of butadiene can rotate to correlate with the lowest occupied MO of bicyclobutane:

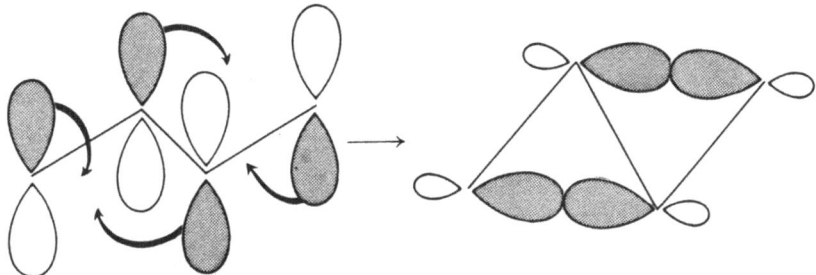

This mode of rotation, however, "pulls" the lowest filled MO of butadiene into an antibonding situation that does *not* correlate with the highest filled MO of bicyclobutane:

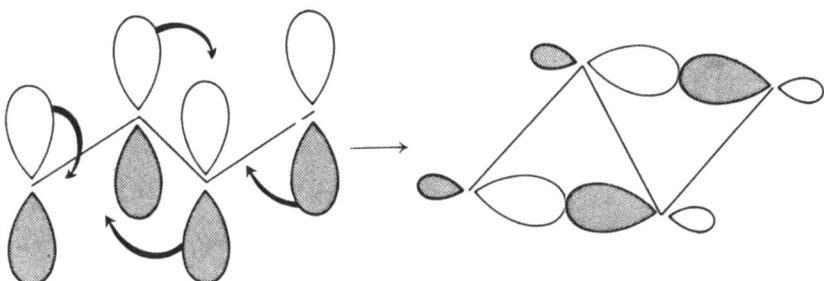

Hence the ring closure is forbidden since *all* filled orbitals must correlate (see discussion of correlation diagrams).

7. Disrotatory (no smooth correlation):

A π_4^* .... σ^* A

S π_3^* ⟋⟍ π^* A

A π_2 ⇅ ⟍⟋ ⇅ π S

S π_1 ⇅ .... ⇅ σ S

Conrotatory (smooth correlation; allowed):

S π_4^* ___   ___ σ^* A

A π_3^* ___   ___ π^* S

S π_2 ⇅ ⟍ ⇅ π A

A π_1 ⇅ ⟍⟋ ⇅ σ S

---

8. Addition of the enolate to the carbonyl may actually occur *faster* than addition at the β carbon. However, the former is a reversible reaction so that eventually all the reactant is converted into the more *stable* 1,5-diketone. Thus, the product composition is not governed by orbital interactions but rather by the thermodynamic stability of the products.

$$R-C\!\!=\!\!CH_2 \longrightarrow \overset{R}{\underset{O}{C}}-CH=CH_2 \rightleftharpoons R-\overset{}{\underset{O}{C}}-CH_2-\overset{R}{\underset{O^-}{C}}-CH=CH_2$$

9. The interaction of the two σ bonds with the *p* orbitals of the carbonyl group apparently gives rise to an unsymmetrical electron density as shown below:

The electron-rich nucleophile thus attacks from the side with the lesser density. This rationale must be considered quite speculative.

10. The undistorted and distorted systems are shown below:

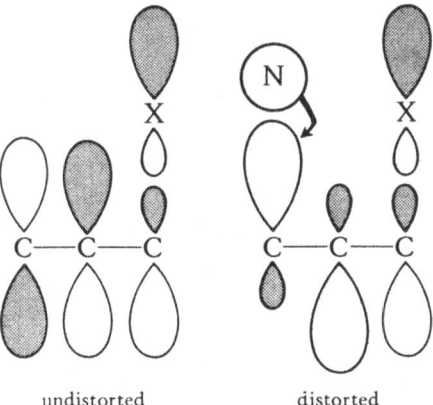

undistorted   distorted

The diagrams incorporate the empty antibonding MO of the alkene (which accepts the nucleophile) and the antibonding σ orbital joining carbon with X. In the undistorted diagram there is no obvious reason why the nucleophile should prefer one side over the other. But in the distorted system the bottom lobe of the central carbon has been enlarged to maximize bonding with the lobe to its right. The leftmost bottom lobe then contracts to minimize antibonding with the enlarged central lobe. If this is an accurate description, then the nucleophile would attack on the *same* side as X because this side has the larger lobe.

---

11. Conrotatory ring opening can give two products depending on whether the relevant carbons both rotate clockwise or both counterclockwise. The left-hand structure is preferred for steric reasons (less nonbonded interactions between the methyls).

---

12. This is a $4n$ system (8 π electrons) which closes in a disrotatory manner when exposed to light:

13. The N of NO_2^+ bonds to the C_1 of naphthalene to give (after expulsion of a proton):

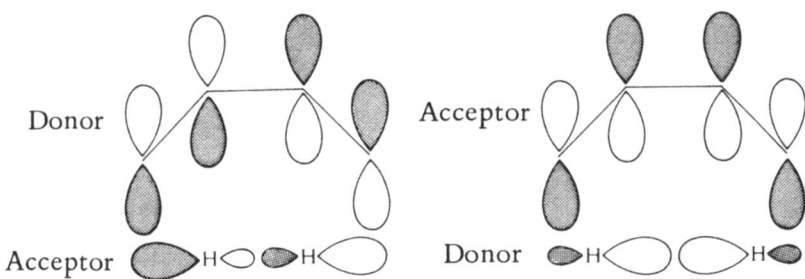

14. It is easiest to consider the addition of H_2 to the diene (if a reaction is allowed in one direction it must be allowed in the other). The 1,4 addition is permitted whether the diene acts as donor or acceptor:

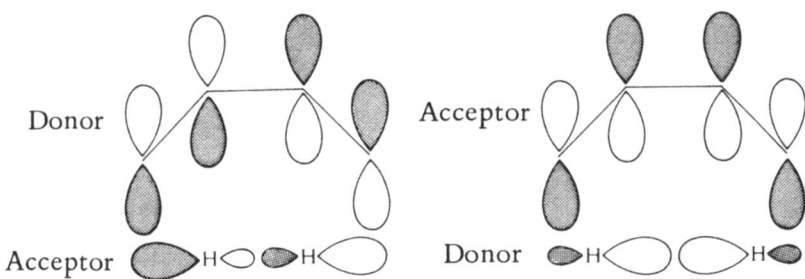

Concerted 1,2 addition is clearly forbidden because of mismatching.

15. This reaction is allowed:

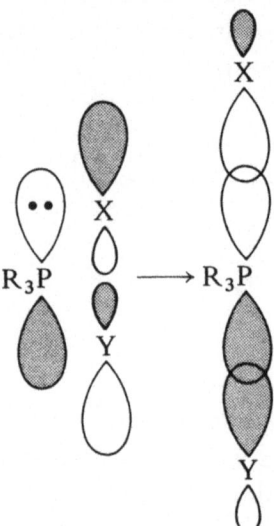

16. Concerted cycloadditions can be achieved by using the lowest unoccupied MO of the ketene and the bonding MO of the ketone:

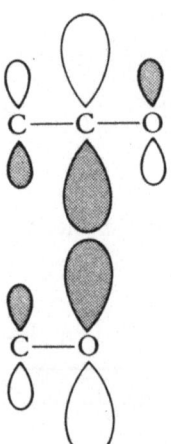

Overlap between the two large lobes gives cyclic ester as the major product.

---

17. Orbital interaction during the 1,5 alkyl shift for the slow compound is permitted:

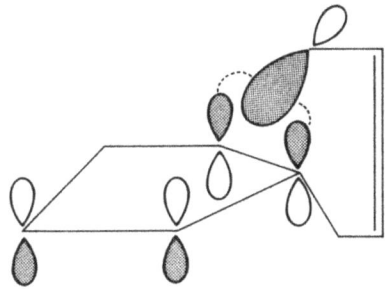

But the vinyl migration is much faster because of an additional ("secondary") interaction between the LUMO of the double bond and a HOMO of the cyclopentadienyl system:

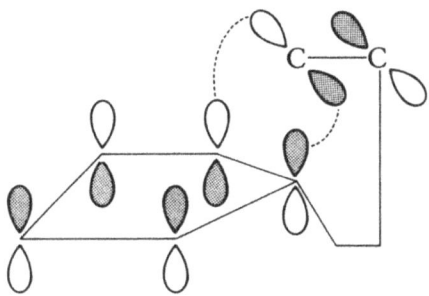

Suggested Reading

1. I. Fleming, *Frontier Orbitals and Organic Chemical Reactions*, John Wiley and Sons, New York (1976).
2. G. Klopman in *Chemical Reactivity and Reaction Paths* (G. Klopman, ed.), John Wiley and Sons, New York (1974).
3. R. G. Pearson, *Symmetry Rules for Chemical Reactions*, John Wiley and Sons, New York (1976).
4. M. J. S. Dewar and R. C. Dougherty, *The PMO Theory of Organic Chemistry*, Plenum Press, New York (1975).
5. A. Liberles, *Introduction to Molecular-Orbital Theory*, Holt, Rinehart and Winston, New York (1966).

Index